HARRY TI

Management of the Hanford Engineer Works in World War II

How the Corps, DuPont and the Metallurgical Laboratory fast tracked the original plutonium works

Published by
ASCE Press
American Society of Civil Engineers
345 East 47th Street
New York, New York 10017-2398

ABSTRACT:

This book presents an exemplary case of successful crisis management. Combining in-depth research, along with first-hand interviews with key participants in the project, the author engagingly recounts an important aspect of America's race to combat the perceived German threat in the development of atomic weapons. He describes the organization and management methods involved in one of the major engineering achievements of the 20th century—the design and construction of the original plutonium-production plant at Hanford, Washington, during World War II. The efforts of the U.S. Army Corps of Engineers, DuPont, and the Metallurgical Laboratory are described in detail, in addition to detailed discussions of the desperate nature of the crisis, labor conditions, technologies, and intangibles of those wartime years. This book also contains information not available in any other published work.

Library of Congress Cataloging-in-Publication Data

Thayer, Harry.
Management of the Hanford Engineer Works in World War II : how the Corps, DuPont, and Metallurgical Laboratory fast tracked the original plutonium works / Harry Thayer.
p.   cm.
Includes bibliographical references and index.
ISBN 0-7844-0160-8 1.
Manhattan Project (U.S.)—Management.   2. Hanford Engineer Works—Management. 3. Plutonium—Metallurgy. 4. United States. Office of Scientific Research and Development. Metallurgical Laboratory—Management. 5. E.I. du Pont de Nemours & Company—Management. I. Title.
QC773.3.U5T43   1996 96-5570
669'.2934'068—dc20   CIP

The material presented in this publication has been prepared in accordance with generally recognized engineering principles and practices, and is for general information only. This information should not be used without first securing competent advice with respect to its suitability for any general or specific application.
The contents of this publication are not intended to be and should not be construed to be a standard of the American Society of Civil Engineers (ASCE) and are not intended for use as a reference in purchase specifications, contracts, regulations, statutes, or any other legal document.
No Reference made in this publication to any specific method, product, process or service constitutes or implies an endorsement, recommendation, or warranty thereof by ASCE.
ASCE makes no representation or warranty of any kind, whether express or implied, concerning the accuracy, completeness, suitability or utility of any information, apparatus, product, or process discussed in this publication, and assumes no liability therefore.
Anyone utilizing this information assumes all liability arising from such use, including but not limited to infringement of any patent or patents.

Photocopies. Authorization to photocopy material for internal or personal use under circumstances not falling within the fair use provisions of the Copyright Act is granted by ASCE to libraries and other users registered with the Copyright Clearance Center (CCC) Transactional Reporting Service, provided that the base fee of $4.00 per article plus $.50 per page is paid directly to CCC, 222 Rosewood Drive, Danvers, MA 01923. The identification for ASCE Books is 0-7844-0160-8/96 $4.00 + $.50 per page. Requests for special permission or bulk copying should be addressed to Permissions & Copyright Dept., ASCE.

Copyright © 1996 by the American Society of Civil Engineers,
All Rights Reserved.
Library of Congress Catalog Card No: 96-5570
ISBN 0-7844-0160-8
Manufactured in the United States of America.

This book was set in Palatino type by JAM Graphics, Emeryville, CA

The editors were Mary Grace Luke and Linda Schonberg

Figures were done by JAM Graphics and Harry Thayer

Photo credits: The Department of Energy has given permission for use of those photos credited herein to the Hanford Science Museum.

DEDICATION

To the engineers and builders of the great projects of the Thirties and Forties, particularly Frank Matthias and Walter Simon, who made major contributions to the Hanford Project and who died shortly after making major contributions to this book.

TABLE OF CONTENTS

Preface — vii

1	Introduction	1
2	Plant Characteristics	5
3	Initial Uncertainties and Difficulties	18
4	The Manhattan Effort	21
5	The DuPont Effort	30
6	The Metallurgical Laboratory Effort	75
7	The Costs of The Hanford Works	81
8	Labor Conditions	89
9	Technologies of The Forties	95
10	Intangibles	98
11	Why Hanford Succeeded A Summary of Lessons Learned	103
12	Afterword	108
13	Acknowledgments	111
14	Endnotes	113

Appendices

A	List of Figures and Tables	122
B	Abbreviations and Acronyms	123
C	Glossary	124
D	Personnel	125
E	References	127
F	Chronology of The Hanford Project	133
G	Calculations	142
H	The Interviews	158

Index — 221

PREFACE

The Manhattan Project was a byword to managers and engineers of the Forties and Fifties. Its mammoth size, its enormous complexity, its first-of-a-kind components, and its mind-boggling schedule were a combination seldom, if ever seen. It was for years a touchstone for engineering-and-construction management performance.

I had picked up these favorable impressions of the Manhattan Project without having any idea of the program's management details, so when the opportunity for first-hand investigation of the topic unexpectedly presented itself, I naturally jumped at the chance. It was an absorbing and thoroughly enjoyable task – the chance to talk to a lot of interesting and authoritative people and the opportunity for several productive trips to archives in Richland, Washington, and Wilmington, Delaware.

These archives have done a thoroughly professional job of cataloging their collections, but they will be the first to agree that Hanford's available documentation is by no means complete. This problem of incomplete records, however, was not a crippling handicap in this study. The combination of authoritative testimony from the high-level management survivors who actually did the job, together with the available Hanford records, gives a surprisingly coherent and complete account of the conduct of the project.

There was additional information that I would have liked, of course, but we must recognize that private companies are not in the library business. DuPont, like every other production-oriented firm, has other goals than thoroughly cataloging and disseminating every Hanford-related document in a hundred brown-cardboard boxes. Still, some of this missing information would be not only of historical interest, but would help to fill in a more-complete picture. I close the Afterword, Chapter 12, with a list of items that would usefully add to the Hanford record.

It is to be hoped that my interpretation of the interviews and of the written Hanford record will trigger discovery of original Hanford data that I did not find, induce discussion of Hanford's management among engineering and construction managers, cause further investigations of the topic along lines that have not occurred to me, and elicit comments on errors that I probably have made, all culminating in a more-complete account of the original Hanford Engineer Works.

Harry Thayer

THE AUTHOR

Harry Thayer holds a BA in Geology – 1950 – from the University of California, Berkeley.

Immediately following graduation, he began a thirty-five year career with Kaiser Engineers in Oakland, California, beginning with four years in construction engineering. This period included four projects: cement and alumina plants; a gasoline refinery; and the twin 100-K plutonium-production reactors at Hanford, Washington.

As a civil engineer in the Design Office he participated in the design of over 70 industrial plants in the areas of cement, iron ore, steel, alumina, aluminum, minerals, mining, nuclear, space, defense, and miscellaneous Government projects.

He was surface-facilities Design Manager for the conceptual designs for the deep geologic repositories for spent nuclear fuel in bedded salt and basalt.

During World War II he was in an Aviation Engineer Battalion, building military airfields in the Pacific Theater.

Chapter 1

INTRODUCTION

This book is an account of how America in the early 1940s organized itself to pull off one of the greatest scientific, engineering, and construction feats of all time.

The Hanford Engineer Works was the plutonium-production component of the Manhattan Project that in World War II made America the world's nuclear leader. It was a mega project on the far edge of totally new and only partially developed sciences and was planned under conditions of the most extreme crisis. Even though there were no precedents whatsoever for its complexities and its exotic components, the exigencies of war had driven it into immediate construction in late March of 1943.

Despite these daunting obstacles, this mammoth process plant of more than 500 buildings worth $4.4 billion in 1994 dollars was completed, and its first product was shipped in under 23 months.

Exactly how did that happen? How did the Manhattan District and its turnkey engineer-constructor, E. I. DuPont de Nemours, produce immediate results despite the near-total uncertainties associated with the project? What management methods enabled such rapid construction? These questions, incompletely addressed in the literature, are the subject of this book.

But first, why is Hanford's management of a half century ago of any real significance, other than as a curiosity of engineering history? The principal answers to that question are these:

- Hanford was an impressive example of the projects of an impressive era in construction.
- Hanford was an example of the best in government Architect/Engineer cooperation.
- It is an advantage simply to know that such spectacular performance is possible under such adverse circumstances.
- It is of interest to determine how the Hanford methods functioned in that crisis, and if they may again prove useful, should similar crises arise in the future.

Information Sources

By extreme good luck, a few surviving, high-level managers who engineered, built, and operated Hanford were still available for interviews, both in person and by telephone. It is their independent but mutually reinforcing testimonies that provided the principal bases for this description of Hanford's management.

In chronological order of their interviews, these men were:

Colonel Franklin T. Matthias.[a] Officer in Charge at Hanford; formerly, General Groves' Deputy Manager of construction for the Pentagon; subsequently managed major civilian heavy-construction projects; retired as Vice President, Heavy Construction, Kaiser Engineers.

Professor Glenn T. Seaborg. Section Chief, Final Products in the Chemistry Division of the Met Lab, where he directed development of plutonium chemistry as the basis for chemical engineering of Hanford's separation process; formerly co-discoverer of plutonium as faculty member of the University of California (UC); subsequently head of the Atomic Energy Commission and Chancellor of University of California; Nobel Laureate, 1951.

Walter O. Simon.[a] Hanford's plant Operations Manager, a member of DuPont's TNX Department; subsequently DuPont's Buffalo Plant Manager.

Raymond P. Genereaux.[a] DuPont's Assistant Design Project Manager for separation-plant engineering; former Head, Chemical Engineering Branch of DuPont's Pure Science Research Group; only five-time contributor to the Chemical Engineers' Handbook.

John B. Tepe.[a] Research chemist loaned by DuPont to the Met Lab for development of the chemical separation process.

Russell C. Stanton.[a] DuPont's Division Engineer in charge of construction of the three reactors at Hanford; former construction engineer on DuPont process projects; subsequently Assistant Field Project Manager for construction of the AEC's Savannah River plant.

Lombard Squires. Head of separation-plant process coordination in the Technical Division of DuPont's TNX Department; former instructor under Lewis in chemical engineering at MIT; subsequently member of the Advisory Committee on Reactor Safeguards of the AEC.

These firsthand accounts were supplemented by Matthias' personal files, DuPont's Design, Procurement, and Construction histories, other unpublished DuPont files, Seaborg's published diary, and the other literature listed in Appendix E.[b]

Mr. Genereaux reviewed the entire manuscript, and Mr. Stanton reviewed the Hanford Section of Chapter 5; their comments have been incorporated into the text.

Content of the Book

Given the passage of 50 years since the Manhattan Project, the terminology and characteristics of the original Hanford Engineer Works (HEW) and the general circumstances of the time may have been forgotten. Familiarity with these matters may be acquired from the chapters on Plant Characteristics, Technologies of The Forties, and Intangibles. On the other hand, those familiar with the original Hanford project may wish to go directly to the chapters on management – The Manhattan

---

[a] The interviews with the men so noted are quite long and are included in full in Appendix H.

[b] This quantity of references and the high concentration per page of facts and concepts extracted from them caused an abnormal frequency of endnotes. One's understanding of the text will not suffer if one wishes to leave these endnotes to future researchers.

Effort, The DuPont Effort, The Metallurgical Laboratory Effort, and Labor Conditions. The principal reasons for Hanford's success are summarized in Chapter 11.

The chapter "Costs of The Hanford Works" provides several measures of management efficiency and other interesting aspects of Hanford costs.

Hanford's uncertainties and difficulties, described in Chapter 3, are two of the principal differences between Hanford and just any large, merely difficult fast-track project. They signify not only the desperate nature of the crisis but also the magnitude and brilliance of the Hanford achievement.

Included in the Appendices are a Hanford chronology, a list of prominent and frequently mentioned personnel, the interviews, and the supporting calculations, as well as the usual report information – the list of figures and tables, the definitions of abbreviations and acronyms, the glossary, and the list of references.

The Entities

Those familiar with America's organization of the war effort may not need to read this section. Other readers, however, will benefit by these descriptions of the entities involved in the atomic program and the relationships among them. In chronological order of participation, they were:

National Defense Research Committee (NDRC). Established by the President at the urging of Vannevar Bush, acting for the National Academy of Sciences. Its goal was to conduct a research program by contracting with universities and private and public institutions, using funds transferred from the Army and Navy.[1] The NDRC subsequently became a subsidiary of the OSRD, described next.

Office of Scientific Research And Development (OSRD). Established by the President to accomplish scientific research for any nation, the defense of which was considered by the President to be vital to the defense of the U. S., and to serve as the liaison office for such research.[2] The NDRC and the OSRD funded and coordinated the initial scientific efforts necessary for national defense, and were founded in June of 1940 and 1941, respectively.

S-1 Committee. The branch of the OSRD responsible for the atomic effort. The successor to the original Uranium Committee established by Roosevelt in response to Einstein's letter explaining nuclear fission.

Metallurgical Laboratory (the Met Lab). The University of Chicago organization that provided the scientific bases and the conceptual designs for the Hanford pile and separation plant. Under contract to the S-1 Committee and later to the Manhattan District, it operated the 1 kW pile at the Argonne experimental laboratory in the Argonne Forest near Chicago and the 250 kW pilot reactor and the pilot separation plant at Clinton.

Manhattan Engineer District (the MED). An ad hoc "district" of the Corps of Engineers comprising the entire U. S., charged with the procurement for, and the design, construction, and operation of, facilities for manufacturing the plutonium and uranium bombs, and the planning for their military application.

Military Policy Committee. A committee set up on the request of the Secretary of War to advise the Secretary and the head of the atomic project on matters concerning that project.

Clinton Engineer Works (the CEW). The area near Knoxville, Tennessee containing the U235 production facilities, the Met Lab's 250 kW pilot, air-cooled plutonium-production reactor, the pilot separation plant, and the SMX non-nuclear experimental pile. Later known as the Oak Ridge Plant.

DuPont. The E. I. DuPont de Nemours chemical company at Wilmington, Del. that, on Groves' urging, accepted responsibility for design, procurement for, and construction of the Hanford Engineer Works and the pilot facilities at Clinton. Operated the Hanford Works to produce plutonium.

Explosives TNX (TNX). The new, ad hoc manufacturing unit of DuPont established to lead and coordinate the design of the HEW, interface with the MED and the Met Lab, and operate the HEW.

Los Alamos. The Manhattan-District facility near Santa Fe, New Mexico that designed and manufactured the nuclear bombs.

Boundaries of The Book

This book addresses the question, and only the question: what engineering, procurement, and construction management methods did the Corps and DuPont employ at Hanford and Wilmington?

It is therefore neither a history of nuclear physics, nor of radio chemistry, nor of the Manhattan Project, nor even of the plutonium branch of that project. It is not intended as a comparison of Hanford management with the management of any specific project, nor with the management practices of any specific entity.

This concentration on a single aspect of the Manhattan Project explains the limited, or nonexistent treatment found herein of other essential activities in the development of the plutonium bomb, e.g.: the Met Lab and its associated Clinton pilot plants; uranium procurement; uranium billet and slug manufacture; and so forth. The only treatment of topics "outside" of the Hanford project, therefore, is that necessary to introduce and frame the topic of Hanford management.

This book is written for engineers. Common engineering terms, concepts, and abbreviations are therefore not explained. Appendices B and C provide definitions of terms not necessarily familiar to all engineers, e.g. "ANSI/ASME NQA1."

Chapter 2

PLANT CHARACTERISTICS

The discussion of Hanford's management is best appreciated with a prior understanding of the plutonium process and facilities of that time, and an idea of the size of the Hanford project, all of which are explained in this chapter.

The process began with the irradiation of uranium by the neutrons obtained from the controlled chain reaction of the pile.

These neutrons converted about 0.025%[1] of the uranium into neptunium. The irradiated uranium – with its contained neptunium – was then discharged from the pile and retained in a storage pool for up to two months for three reasons: to permit a percentage of the neptunium to decay into plutonium; to permit the radioactivity to decrease, thus reducing the handling hazard in the separation plant; and to reduce the radioactivity of Iodine 131 to as low a value as possible – approximately a 10 half- life reduction – to minimize the hazard emanating from the gases discharged from the stacks.[2]

The plutonium-bearing uranium was then chemically treated in the separation plant to extract and concentrate the plutonium. The depleted uranium and the waste fission products were then sent to waste storage.

The 100 Areas And The Piles

There were three of these areas – Areas 100B, 100D, and 100F – on the right bank of the Columbia River, located as shown in Figure 2.1. Each area had a pile building – the 105 building – and a number of support buildings.

The pile, built on a 23 ft deep (7 m) concrete foundation, consisted of a graphite cuboid 36 ft wide (10.6 m) by 36 ft high by 28 ft (8.5 m) in the direction of the fuel tubes, and comprised more than 100,000 stacked graphite blocks, most of which were $4^{3}/_{16}$ in. square (10.6 cm) by 48 in. long (1.2 m). About one quarter of the blocks were drilled with a longitudinal hole to accept a 1.73 in. OD (4.4 cm) by 1.61 in. ID (4.1 cm) aluminum fuel tube, of which there were 2004. The fuel tube was cooled by passing about 14 GPM (53.1 LPM) of water through it. All blocks in the fuel tube layers were beveled on the four longitudinal corners to provide passageways for the helium used for air purging and for leak detection.

Reaction physics dictated an $8\ ^{3}/_{8}$ in. (21.3 cm) fuel lattice spacing, which was attained by spacing the blocks as shown in Figure 2.2.

Each tube penetrated the pile from front to back and contained 32 uranium slugs about 1.5 in. (3.8 cm) in diameter by 8 in. (20.3 cm) long canned in aluminum jackets. Inert, dummy slugs were used to space the fuel to dimensions dictated by reaction physics. The irradiated slugs were pushed out of the ends of the tubes into a

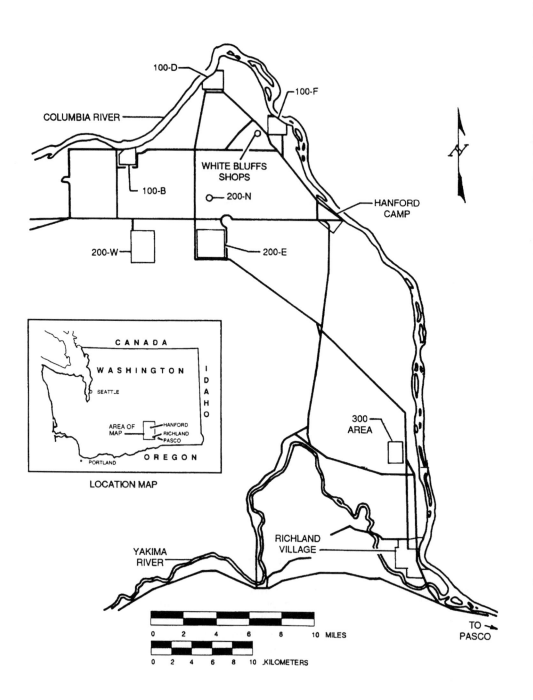

Figure 2.1
THE HANFORD ENGINEER WORKS

Reference: End Note 3

Figure 2.2

BLOCK SPACING

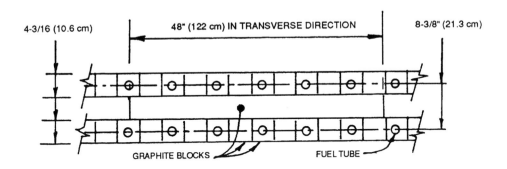

pool of shielding water – then transported to a storage pool in the 212 building before transfer to the separation plant.

The pile was enclosed by two shielding layers: an inner, 10 in. thick (25.4 cm) thermal shield of cast-iron blocks containing cooling water passages, and an outer biological shield consisting of an assemblage of composite, 4 ft (1.2 m) (approximately) cuboids comprised of alternating layers of thick steel plates and bundles of Masonite sheets. This outside layer was provided with flexible seals along the joints – then welded together to form a gas-tight enclosure around the pile. The rest of the 105 building was further shielded from the pile by a thick concrete wall around the pile.

Each fuel tube penetrated the biological shield through a surrounding sleeve 7.5 ft (2.3 m) long, called the gun barrel tube, which in turn penetrated the corresponding tube in the biological shield block, and was supported in the shield tube by a cast-iron shielding sleeve, called a doughnut. A tube nozzle assembly was fitted to each end of the fuel tube and functioned to admit, control, and discharge cooling water, and to admit and discharge fuel slugs. The joint between the gun barrel tube and the outside of the biological shield was made gas tight by an expansion bellows welded to the outside shield plate.

The pile was provided with nine horizontal and movable boron coated control rods driven by hydraulic actuators. These rods functioned to absorb neutrons when it was desired to adjust or stop the chain reaction. Twenty-nine vertical, gravity powered safety rods were provided as emergency backup. Finally, arrangements were made for emergency flooding of the pile with a borax solution through thimbles penetrating the pile.

A slug charging machine was mounted on the elevator that spanned the front face of the pile. A back face elevator contained a tube puller/cutter and a shielded cab for manual performance of various maintenance operations.

Auxiliary systems were provided for supplying cooling water and helium.[4]

A dramatic incident in the development of the pile was an interesting example of the conservative approach used by engineers to compensate for unknowns. This approach proved to be particularly necessary in the new field of reactor design.

BOTH PHOTOS: HANFORD SCIENCE MUSEUM

ABOVE: 100 B Area. Reactor building is between the two water towers. Large building on the left is the High-Lift reactor cooling water pump station.

LEFT: 105 B Reactor building.
(Post-war photo)

BOTH PHOTOS: HANFORD SCIENCE MUSEUM

ABOVE: Reactor charging end. Shows 2004 water and fuel inlet nozzles and water-supply manifolding.

LEFT: Slug-storage pool. (212 Building)

The Met Lab had conceived 1,500 fuel tubes in the central, cylindrical part of the pile, as shown diagramatically in Figure 2.3.

TNX's George Graves requested an extra 504 tubes to fill in the corners. The space was available, so why not use it? The Met Lab physicists argued strongly against TNX's position as being an unnecessary waste of time and money, but DuPont held firm until the Met Lab agreed to the extra tubes.

Figure 2.3

DIAGRAMATIC PILE CROSS SECTION

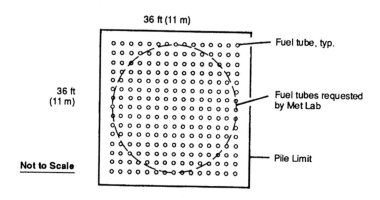

In the initial reactor loading, however, fuel slugs were placed only in the central, cylindrical core. Shortly after B Reactor startup and initial criticality, a fission product, Xenon, had built up in the cans until it poisoned the reactor sufficiently to bring the nuclear chain reaction to a complete stop. After Wheeler, Fermi, and Greenwalt deduced the cause, it turned out that the solution was to place fuel slugs in the extra tubes outside the core. The resulting decrease in neutron leakage overwhelmed the poisoning effect and the reactor worked.[5]

For related aspects of this incident, see page 53.

The CMX Unit At Hanford

This experimental installation of six full-size fuel tubes, with auxiliary pumping and water treatment facilities, was built and operated to study the corrosion effects of Columbia River water on the tubes, and the water treatment chemicals required.[6]

The SMX Unit At Clinton

This was a full-scale, non-nuclear section of the proposed Hanford pile constructed to study design and construction features of the Hanford pile, e.g.: proving of graphite machining; analyzing laying procedures; investigating expansion effects; studying pile deformation under load; and other matters.[7]

The 200 Areas - The Separation Plants

Two separation areas – 200 East and 200 West, located about 4 mi (6.4 km) south of 100-B (Figure 2.1) – contained three separation plants, two at 200 West and one at 200 East. Each separation plant consisted of a process cell building, or "canyon" (221 Building), a concentration building (224), a purification building (231), and magazine storage (213). The three slug storage pools (212) were at 200 North. The separation process performed in these buildings recovered 90% of the available plutonium,[8] and is shown in Table 2.1.

The Canyons (221)

Each canyon building, over 800 ft (244 m) by 65 ft (19.8 m) wide, consisted of a line of 40 process cells, each 17 ft 8 ins. (5.4 m) by 13 ft (4 m) by 20 ft (6.1 m) high, surrounded by thick concrete shielding and a removable concrete cover. (One of the canyons at 200 West was provided with an additional two cells for use as a semiworks.) Operating, electrical, and piping galleries paralleled this cell line, and an overhead crane gallery the length of the building provided for remote and shielded removal of the cell covers and maintenance of the cell equipment and piping. Special tools were provided to permit the crane operators to perform this remote maintenance. Details of design and equipment are provided in Chapter 5.

The Concentration Building (224)

Because the cell operations had greatly reduced the radioactivity of the process stream by removing the fission products (the decontamination steps), a separate building with much less shielding was provided for the concentration steps. Normal (non remote) maintenance was permitted here because of the reduced radioactivity. The process equipment, however, was of the same types as in the canyon, and provided for a concentration ratio of 375:1

The Purification Building (231) And Storage Magazines (213)

One purification building at 200 West sufficed for purification of product from both West and East areas. It used the same types of equipment as for concentration, but at a much smaller, laboratory scale. Because of the extreme toxicity of concentrated plutonium salts, this building was provided with five separate air conditioning systems. The 213 buildings were standard munitions magazines.[11]

Other Areas

The 300 Area, a little north of the Richland Village, contained the plant laboratories, material test facilities including a small experimental and testing pile, a small separation semiworks, and the slug manufacturing and canning facilities.[12]

Neither the construction of this area nor of the very extensive conventional facilities – roads, railroads, power lines, water supply, and the Richland Village – presented any unusual problems, nor were they on the critical path. They are therefore not discussed further in this report.

Table 2.1

SEPARATION PROCESS SEQUENCE

SEQUENCE NUMBER	PROCESS STEP	BUILDING NUMBER
1.	Irradiated Uranium Storage	212
2.	Can Removal	221
3.	Uranium Dissolving	221
4.	Storage of Uranium Solution	221
5.	Extraction	221
6.	First Oxidized Decontamination Step – Bismuth, Ceric and Zirconium Scavenging	221
7.	First Reduced Decontamination Step – Fluosilicate Added	221
8.	Second Oxidized Decontamination Step Bismuth Scavenging	221
9.	Second Reduced Decontamination Step Fluosilicate Added	221
10.	Disposal of Can Solution Waste	241
11.	Disposal of Uranium Solution Waste	241
12.	Disposal of Decontamination Waste	241
13.	Bismuth Oxidized Crossover Step	224
14.	Lanthanum Fluoride Oxidized Crossover Step	224
15.	Lanthanum Fluoride Reduced Crossover Step	224
16.	Metathesis - $KOH \cdot K_2CO_3$ Type	224
17.	Disposal of 224 Building Wastes	224
18.	First Peroxide Cycle	231
19.	Second Peroxide Cycle	231
20.	Preparation of Solution For Shipments[a]	231
21.	Treatment of Solutions Recycled to 224 Bldg.	231
22.	Magazine Storage	213

---

[a] The final product shipped to Los Alamos was plutonium nitrate, originally conceived as a dilute solution. This was found unsuitable for shipment however, as it radiolyzed, producing undesirable gases. It was found that concentrated nitric acid produced a glassy solid.[10]

Reference: End Note 9

ABOVE: LAWRENCE BERKELEY LABORATORY/BATTELLE PRESS/THE PLUTONIUM STORY
BELOW: HANFORD SCIENCE MUSEUM

ABOVE: Canyon foundation-wall construction. Wall forms being placed. A number of process-cell forms are in place, and additional cell forms are seen in the form yard. Rail-mounted traveling cranes each side of excavation. Batch plant, cement-car track and aggregate-car track at top of photo.

LEFT: Canyon end-wall pour. Note crane-hoisted concrete bucket in lieu of pumpcrete rig.

The 221 T "canyon" building, the initial and largest separation building. The tall, narrow protruberances on the long wall are stairwells to the operating areas. The large structure on the long wall of the canyon is the 271 building, used for chemical storage and preparation, shops, ventilation equipment, and supervising offices. (Gerber B, p.14)

The two-level structure behind the canyon is the 224 concentration, or bulk reduction building. The construction batch plant with aggregate and cement rail cars is seen in the left center and foreground.

The 221-T building, opposite side from above photo. Exhauster building and stack, with underground duct in foreground. 224 building across from the far end of 221 building, with 222 sample-preparation laboratory between the stack and 224.

BOTH PHOTOS: HANFORD SCIENCE MUSEUM

In this separation cell in the 221 building, the precipitate is centrifuged from its mother liquor. The top of the motor for the 40 inch (102 cm), solid-bowl centrifuge is labeled 7-2. The motor is two-speed, 40/20 HP (30/15KW), with the 40 HP used for centrifuging. The 20 HP speed is for rotation of the centrifuge to remove the accumulated solids from the inside of the centrifuge by remotely actuated hydraulic plows.

The large vessel, labeled 7-3, is the effluent tank for accumulation of clear liquor from the centrifuge.

The pipe segments - "jumpers" - are provided with remotely operated connectors on each end, which were operated from the overhead crane by means of a suspended impact wrench. This wrench engaged the pointed, black, hex nuts, which then moved the connector jaws.

The rectangular objects near the centrifuge motor are structural supports for the centrifuge, piping, and electrical connections.

PROCESS DATA: RAYMOND GENEREAUX & WATSON WARRINER
PHOTO: LAWRENCE BERKELEY LABORATORY/BATTELLE PRESS/THE PLUTONIUM STORY

Size of The Hanford Works

The project was very large — one of the major engineering and construction works of the twentieth century. (See Table 2.2).

The area of the project was more than 60% of the land area of Rhode Island. The cooling water treatment plants were capable of supplying domestic water for 1.3 million people. Peak employment was eight times that of Grand Coulee Dam and 95% of the Panama Canal. The construction camp was the third most populous city in Washington.[13] The intra plant bus fleet exceeded Chicago's, then the world's largest, by 13%.[14] Bus passenger miles during construction could have provided each of the 45,000 peak work force with 1.3 round trips between San Francisco and New York City.

The total excavation equaled 25% of the volume of Fort Peck Dam[15] and 10% of the Panama Canal. I was at first quite skeptical of a mere process plant, no matter how large, having earthwork of this magnitude, particularly since the identical quantity was reported by several sources, as if everyone had copied the same error. I therefore made a conservative takeoff based on known dimensions of major buildings, the sizes of the various process areas, the lengths of roads and railroads shown on Figure 2.1, the general slope of the terrain that I remembered from my years at Hanford, and minimal road and railroad cross sections.

Incorporating these bases with certain conservative assumptions, I arrived at an estimated earthwork quantity of 72% of the reported earthwork volume, a quantity that I therefore no longer doubt. It was indeed a very large process plant!

Table 2.2

Hanford Works Quantities

	1945 Millions of Dollars		1994 Millions of Dollars
Cost, without uranium charge...............	$333.7		$4,130[a]
Cost, including first uranium charge...........	$357.7		$4,370[a]
	English Units	*Metric Units*	
Land Area: 505 sq mi of the works............ plus 165 sq mi buffer zone (1308 km², 427 km²)	670 sq mi[16]	(1735 km²)	
Process buildings.........................	195[17]		
Support buildings.........................	345[18]		
Water treatment plants capacity.............	129,600,000 gpd[19]	(490,600,000 lpd)	
Concrete, process and support bldgs..........	719,000 cu yd[20]	(550,000 m³)	
Concrete, other	61,000 cu yd[21]	(47,000 m³)	
Lumber, temporary and permanent............	160,000,000 fbm[22]	(378,000 m³)	
Structural steel, temp. and permanent.........	583,000 tons[23]	(529,000 tonnes)	
Pipe, 12 in. dia and under, water............. (30. cm)	232 mi[24]	(373 km)	
Pipe, 24 to 42 in. dia, water	20 mi[25]	(32 km)	
230 kva transmission lines	52 mi[26]		
Permanent highway	386 mi[27]	(621 km)	
Permanent railroad........................	158 mi[28]	(254 km)	
Excavation...............................	25,000,000 cu yd[29]	(19,000,000 m³)	
Carloads of material used	40,000[30]		
Total project manhours	126,265,652[31]		
Peak employment.........................	45,000[32]		
Passenger miles, intra plant buses	340,000,000[33]		

---

[a] See App. G for escalation

Chapter 3

Initial Uncertainties And Difficulties

Those in 1942 with the responsibility for managing the project faced a bleak outlook. Groves put it as well as anyone: "Never in history has anyone embarking on an important undertaking had so little certainty about how to proceed as we had then."

DuPont's Harrington and Stine initially expressed the opinion that the entire project was beyond human capabilities. Even after DuPont had reluctantly agreed to do the work, Stine, agreeing that it was possible, said to Compton, "We must tell you that in our judgment there is not more than one chance in a hundred that it can lead to anything in this war."

The following lists of uncertainties and difficulties were truly sobering, even for the best minds in the country. Seaborg remarked in February, 1942 after scoping discussions with other Met Lab scientists, "As a result of these meetings, I now fully realize the enormity of the chemical separation problem of isolating ^{239}Pu from large amounts of uranium and almost fantastic intensities of fission products." Both the Corps and James Bryant Conant expressed grave doubts of the project's feasibility, particularly with respect to the ability to produce the material in time for effective military application.

Nevertheless, the project had to go forward. Our fear of the German scientists was very real and drove us to desperation measures. An indication of our desperation was Groves telling the Met Lab on October 12, 1942 that there is no objection to a wrong decision with quick results. If there is a choice, he said, between two methods, one of which is good, and the other looks promising, then build both.[1]

Uncertainties

- When the Corps was directed on June 18, 1942 to assume responsibility for direction of the bomb program, there was no experimental proof that plutonium could be produced in a pile. This proof did not come for five months, two months after the design of the production plant had begun.
- The recovery of plutonium from a highly radioactive medium had not been demonstrated. The recovery process was not finalized until more than two years later, two months before separation cell equipment was scheduled for installation.
- Groves told the DuPont Executive Committee prior to securing their agreement to undertake the work that the project might fail, but nevertheless the work must start immediately.

- We had no sure knowledge of Germany's progress towards the bomb, but we were sure they were working on it, and most probably ahead of us.
- In pointing out the serious conflict between the demanding schedule and the near total absence of scientific and engineering information, Groves said, "We could not afford to wait to be sure of anything. The great risks involved in designing, constructing, and operating plants such as these without extensive laboratory research and semi works experience simply had to be accepted."
- Nuclear physics and the handling of radioactive products were entirely foreign to DuPont's experience.
- It was considered possible that, once the chain reaction was started, it could progressively increase to a point above which it would get out of control, resulting in catastrophe.
- DuPont had required a total of 25 years to get nylon and neoprene into mass production, yet these were simple compared to the new plutonium concept.
- The engineering difficulties of the project were considered to be unprecedented.
- There was no advance assurance that all of the unprecedented processes and components to be designed and fabricated in parallel would work, nor, if they worked, would be in time to support each other.[2]

Difficulties

Although it was conceivable that the following individual processes and components of the plutonium program could somehow be actualized, the difficulties en route to completion had never before been encountered.

- Separation process conceptual design proceeded for 15 months with ultra microchemical methods, using invisible microgram quantities of plutonium salts because macro amounts of plutonium metal did not exist, and would not until four and one half months before installation of process equipment was scheduled.
- Final design of the separation cell structure and its embedded piping had to be completed three months before the type of separation method had been decided, and fifteen months before the process design was completed.
- All of the remote handling devices for both the pile and separation plant were without precedent and therefore had to be conceived, designed, and developed from scratch.
- Remote replacement of separation cell components demanded vessel and piping fabrication to hundredths of an inch, requiring painstaking design, mock ups to prove the design, and unprecedented vendor monitoring methods and construction methods.
- A large number of key components were without industrial process precedent, either in design or in fabrication. These included:
  — The eight ton, composite biological shield blocks, the components and overall dimensions of which were to be fabricated to thousandths of an inch.
  — The manufacture and machining to thousandths of an inch of the pile's graphite blocks.

- The aluminum fuel tubes, designed to a fine balance between wall thickness of sufficient strength without undesirable k value reduction. The development and extrusion of these tubes of special interior cross section required eleven months.
- Fabrication of uranium into billets and machining of these into slugs; the metal forming characteristics of uranium had to be developed from scratch.
- Development of uranium slug canning began in March of 1943, but was not completed until mid 1944, a month before the slugs were required for charging B Reactor.[3]

• Finding manufacturers and developing methods for producing graphite and uranium of purity suitable for the pile process.[4]

Chapter 4

THE MANHATTAN EFFORT

Vannevar Bush and James B. Conant had persuaded Roosevelt to form two wartime scientific development agencies – the National Defense Research Committee in 1940 and the Office of Scientific Research and Development in 1941; these agencies encompassed the atomic effort. A year later Bush realized that the scope and magnitude of the atomic effort was far beyond the capabilities of these two university based organizations.

He therefore recommended to Roosevelt in March, 1942 that the Corps of Engineers be given charge of construction of the atomic project.[1]

Groves wrote of this Bush concept as follows: "It is to their everlasting credit that Bush and his colleagues had the discernment to recognize the limitations of their own organization as well as the moral fortitude to admit them in the national interest. Very few men confronted with a similar situation would have done so." [2]

Remarking on the same circumstance, Matthias said, "If Vannevar Bush hadn't forced them to put all this stuff together under one control, we would never have got it done in time."[3]

The MED was formed in June, 1942 and Groves was placed in charge in September.[4]

The National Organization And Activities

A simplified national organization chart is shown in Figure 4.1. This is a summary of the complete organization, which in its totality encompassed 15 headquarters departments and 14 branch offices throughout the nation.

Figure 4.2 isolates the Hanford-related command and administrative lines. (The apparent conflict between the two charts in the Nichols/Hanford relationship is explained below under the topic "The Hanford Organization.") The principal feature of this figure is the direct line of authority from the President to Hanford in three short steps. That, to Frank Matthias, was one of the most important organizational aspects of the Hanford job. He commented: *"We had one authority! That made the difference. I want to strongly emphasize that. It was one of the things that made it successful."*[6]

The straight line of authority was simplified and expedited by General Reybold's ad hoc, voluntary withdrawal from the chain of command. (He had been Groves' commanding officer.) As Matthias said: "General Reybold recognized the intent of the orders to General Groves and offered his help without exercising authority . . . This attitude of a Corps Commander, with respect to actions of one of his officers, was probably without precedent."[7]

Another major simplification was the naming of Groves to the position of Executive Secretary of the Military Policy Committee in September of 1942.[8] This

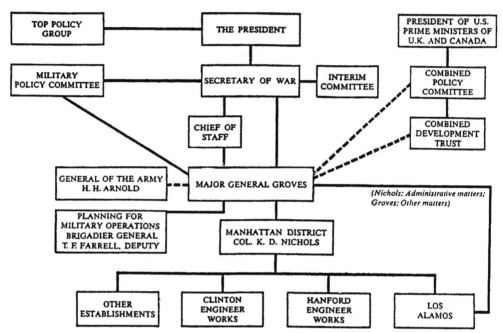

Organization of Atomic Project, May, 1945 (Simplified chart)

Figure 4.1

MANHATTAN PROJECT
SIMPLIFIED ORGANIZATION CHART

Reference: End Note

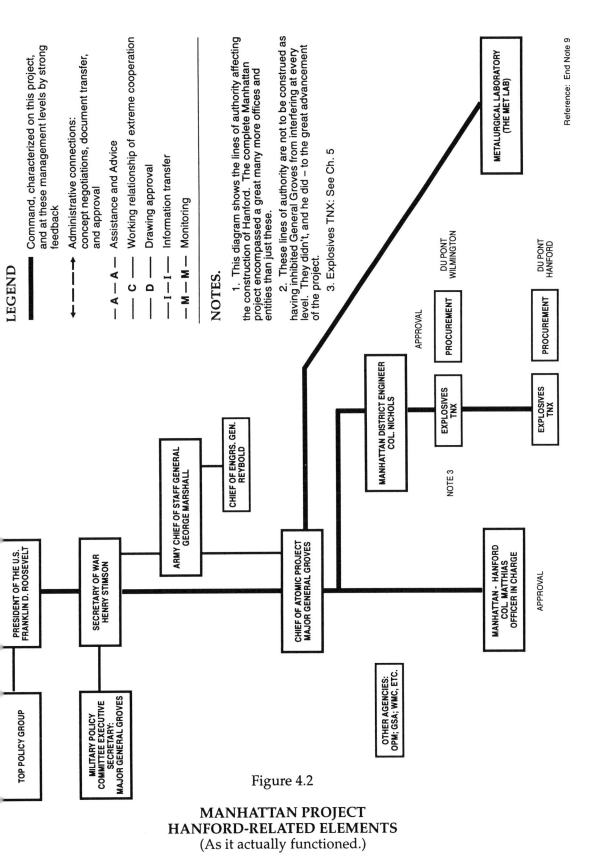

Figure 4.2

**MANHATTAN PROJECT
HANFORD-RELATED ELEMENTS**
(As it actually functioned.)

Colonel Franklin T. Matthias

Major General Leslie Groves

BOTH PHOTOS: HANFORD SCIENCE MUSEUM

expanded his scope from simply the head of nuclear plant construction to the Executive for the entire atomic program.

The balance of Figure 4.2 is discussed under "The Hanford Organization" below, and in Chapter 6.

In addition to the above characteristics, a number of other MED activities and attributes contributed to Hanford's rapid completion; they are described next.

- **Priority.** Groves' first move after appointment as head of the MED was to call on Donald Nelson, head of the War Production Board, to obtain the highest wartime priority – AAA – for the Manhattan Project, thus correcting a deficiency that had seriously delayed the project.[10]
- **Authority and responsibility** were kept together
- **The command channels** were totally understood by all involved in them, notwithstanding the major exception described below under "The Hanford Organization And Activities."
- **The internal organization** was simple and direct, enabling Groves to make fast, positive decisions. He eschewed large staffs as sources of inaction and delay.[11]
- **On-the-spot decisions.** Nichols words: "From the start, our philosophy was to go where the work was being done and make decisions as needed on the spot."[12]
- **Verbal authorization.** In Nichols' words: "To save paperwork and time, as well as to preserve secrecy, very little written material passed between Groves and the District."

  "He asked if I wanted more in writing, or to continue with just verbal instructions and authorizations. I told him that so long as I participated in the preparation of progress reports to the President, or saw the approved reports, verbal instructions were adequate. I did not see how it would be possible to cover in writing all the informal decisions that were made at the numerous meetings involved. Formal procedures would be time consuming for both of us and definitely would slow the work."[13, a]

- **Financial control.** It was remarkable that two billion then year dollars ($27 billion, 1994) were expended in fourteen locations scattered across the nation in a 2.8 year crash program without loss of financial control. Nichols described how that worked: "The district set up its own finance officers and an internal auditing staff. We maintained amicable relations with the comptroller general of the U. S., primarily because we mutually agreed to establish auditors at CEW and HEW *to keep the auditing current* (emphasis added). As a result the comptroller auditor completed his audit within thirty days from the time we had expended the money. My administrative people were very pleased when the Comptroller General reported to the Senate Special Committee on Atomic Energy in April, 1946:

  'We have audited, or are auditing, every single penny expended on this project. . . . I might say it has been a remarkably clean expenditure . . . The very fact that our men were there where the agents of the Government

---

[a] It should be emphasized that the effective verbal shortcuts discussed above were communications between Nichols and Groves, and between Groves and his other immediate subordinates. As will be made clear in subsequent discussions of Manhattan at Hanford and of the DuPont effort, all of the necessary and routine paperwork for engineering, procurement, and construction were completed in minute and rigourous detail.

could consult with them time after time assured, in my opinion, a proper accountability . . .' I believe this achievement occurred because of the way in which Colonel Marshall started the project and organized the administration of the district."14, a

- **Congressional oversight:** There was none. In today's terminology, Manhattan was a black program.
- **Legal controls.** This topic is beyond the scope of this book. Suffice it to say that both the District and DuPont meticulously covered every potential legal problem, even to the point of DuPont's request for Government documents giving the Government's bases for Nichols' and Groves' authority to contract with DuPont! Those interested might consult Nichols, pp. 82, 83, and 132 and the DuPont documents mentioned in App.H 1, 26.FM
- **Immediate site investigation and decision.** A Corps/DuPont team provided an efficient and rapid site investigation in the last half of December, 1942, leading to final selection 16 days after the start of the site trip. In that time the site team considered eleven sites in four states in areas near the following locations:

    Idaho: Coeur d'Alene
    Washington: Northwest of Grand Coulee Dam; Mansfield; Moses Lake; Hanford; Horse Heaven northeast of Plymouth.
    Oregon: Deschutes River/Madras area; John Day River.
    California: Pit River; Needles; Blythe.

The site team wrote its trip report on the 19 hour return flight to Washington, D.C. and delivered it to Groves on January 1, 1943. Groves, General Robbins, the Deputy Chief of Engineers, and Carl Giroux, the Corps' power expert, concurred on Hanford at that meeting.b

A principal reason for this rapid determination, besides the experience and expertise of the site team, was that the sole authority for site approval was vested in the MED. No other agencies – Federal, state, or local – had jurisdiction.

The significance of this very early site selection was that it enabled an early start on such critical-path activities as: plant layout; relocation of existing military and civilian facilities and activities; soils exploration; power and telephone arrangements; camp and commissary planning; and warehousing arrangements.

The above attributes and activities help to explain the small size of Groves' headquarters office. He began the project with two rooms and ended three years later with seven. (In the last month before August 6 he added additional space to cope with the public relations task connected with the bomb drops.15)

The Hanford Organization And Activities

In Figure 4.2 it is seen that Matthias reported directly to Groves, unlike Figure 4.1. This arrangement was because Nichols was fully occupied in managing both the national program and in the direct management of Clinton. In his book, Nichols said that he got to Hanford only about four times a year. He exercised administrative control, however, over the entire program.

---

a Colonel Marshal of the Corps Engineers was Colonel Nichols' predecessor as Manhattan district engineer.

b A full account of this site investigation trip is written up in Col. Matthias' interview, App. H-1, 1-16.FM. Civil engineers should find it of considerable interest.

Matthias remarked on his relationship with Nichols: "It is a great tribute to the quality of Col. Nichols that no friction developed as a result of this unusual organizational relationship, and I enjoyed his complete support."[16]

Of the greatest importance to Hanford's spectacular construction schedule was Matthias' complete authority for approval of DuPont's field procurement of $2.6 billion (1994 dollars) of material, equipment, and subcontracts. He was in fourth position below the President in procurement authority. "I didn't have to go to anybody to get approval," Matthias said.[17] *The Defense Acquisition Policy, to use today's terminology, consisted in its entirety of buying whatever DuPont and Matthias determined to be necessary, enormously accelerating Hanford's construction.*

The Hanford Organization Chart, Figure 4.3, is a simplified version of Matthias' complete 1945, nine-page chart, and displays the principal functions of the Officer In Charge.

In addition to managing the 500 Manhattan personnel at Hanford, the OIC's major tasks were monitoring DuPont's scheduling and construction activities, procurement, costs, and payroll. As both Matthias and Nichols pointed out, Hanford had resident Government auditors, in addition to the OIC's auditors to ensure correct DuPont financial reports before they ever left Hanford.

Other tasks included: construction equipment procurement through the Corps' Washington, D.C. headquarters; expediting; priorities; labor relations[a] and recruiting; and monitoring DuPont's plutonium production.

A major task that fell on Matthias personally was relations with various governmental entities – local, state, and national – as he remarked in his interview.

"I established a working relationship with the Governor and several of the wartime agencies – Federal usually – and I got well acquainted with the Congressmen and Senators, and I didn't have any problems with them. The Congressman was Hal Holmes – a big help to me. The Governor was wonderful – he assigned one of his key people to take care of any problems we had, and I could call him. (He didn't know what the project was for.)"

"We worked out legal things with Benton County. And when they turned up, we'd find some way to take care of it. It was never a problem. With respect to County and State building inspectors," Matthias remarked, "They didn't come in. But we had a working agreement with them and we followed their rules."

"You know, we had a remarkable amount of cooperation, and I think it was the DuPont people who were reasonable and they didn't try to pull any wise stuff on them, and I had good relationships with them."[19]

With respect to expediting, Matthias pointed out that there were no expediting problems because they didn't wait for problems to develop. "For important items we would put an officer or other expediter in the factory, and he'd get right after them to make sure they took care of our needs. They tracked rail shipments, keeping after the railroads until the shipment arrived. I never heard of its being a problem," Matthias said.[20] (Engineering and construction managers will find this final sentence of Matthias' astounding for either a wartime or peacetime project!)

Concerning Matthias' communications with his staff, he remarked that it was "almost 100 percent oral." "I remember once in a while writing them something." And he held no regularly scheduled progress review meetings, or any other kind. "I didn't have any standard meetings; I'd call a meeting when something was necessary."[21]

The OIC DuPont relationship was one of close cooperation. Stanton remarked, "We had total support from the Corps."[22] This was by intent. Matthias observed,

---

[a] The OIC's labor-relations effort is described in Chapter 8.

Figure 4.3

MANHATTAN/HANFORD
SIMPLIFIED ORGANIZATION CHART

This chart shows the branches that encompass the principal Officer-In-Charge (OIC) functions. Subsidiary functions are omitted.

- **OFFICER IN CHARGE — Franklin T. Matthias**
  - **Chief of Construction — Lt. Col. Benjamin T. Rogers**
    - Monitor DuPont's Design, Drawings, & Drawing Schedule
    - OICs for 100, 200, & 300 Areas (Review & reporting of construction)
    - Expediting & Priorities (Control Section)
    - Equipment Procurement Section
  - Public Relations Control
  - **Executive Officer — Major William Sapper**
    - Labor Relations & Recruiting
    - "Surveying". Disposition of Government Property
    - Engineering & Maintenance Divison
      - Communications — Operations & Maintenance
      - Community Management Branch
        - Real Estate Acquisitions
        - Bids, Contracts, Records, Leases, Rental Rates
      - Engineering Branch
        - Coordination of Building Maintenance with DuPont
        - Electrical Power: Coordination with Power Companies, Maintenance
        - Mechanical: Maintenance of Barracks, Boilers, Water Supply
        - Civil: Roads & Site Work
  - **Administrative Division — Major H. D. Riley**
    - Fiscal Branch
      - Cost Review — Construction & Production
      - Finance Section — Accounts Payable
      - Audit Section — Payroll, Purchase Orders, Invoices, Subcontracts
      - Civilian Personnel Section
    - Property & Transportation Branch
      - Dispatch and Repair of Corps Vehicles
      - Accounts & Records Sub Section
      - Audit DuPont Property Records
      - Equipment & Field audit Sub Section
      - Air & Water Patrols
      - Procurement Section: Review DuPont Purchase Orders
  - **Production Division — Major J. F. Sally**
    - Monitor DuPont's Production & Production Records

Reference: End Note 1

with respect to Gil Church, the DuPont Construction Manager, "We two 34 year olds promptly established a working, team relationship that pervaded our developing organizations, dedicated to the single objective of fast and efficient construction of the project."[23]

Top level communications between the OIC and DuPont were oral, possible said Matthias because he and Gil Church "had an unusual respect going." Matthias said that he never wrote confirming memos to DuPont. "Everybody trusted each other on the management side." As a further time saver, Matthias read DuPont's internal reports rather than requiring written reports to the OIC. He held no scheduled meetings with DuPont, but talked to them whenever he needed to.[24]

The OIC— Washington, D.C. relationship was one of mutual confidence. For the most part, Groves gave Matthias a free hand at Hanford. Matthias did however, have to defuse a potential friction point – Groves' constant short circuiting of the chain of command. As Nichols observed, "Groves would regularly contact directly for information, or to issue instructions, any officer in the Manhattan District or any individual working for our various contractors for research, engineering, construction, or operations whenever he needed to. Organizational channels meant nothing to Groves."[25]

Matthias took the broad view of this interference; he realized that it would only hinder the prosecution of the job if he protested to Groves. "I just told my people to talk to Groves whenever he contacted them," he said, "and to write down everything he said – then come and tell me about it."[26]

Chapter 5

THE DU PONT EFFORT

DuPont entered the plutonium project in a relatively minor way, with no idea of the central role it would ultimately play. The initial contact, on July 14, 1942, was from the Met Lab's J. G. Stearns, Director of the Argonne Forest pilot plant. He requested that DuPont loan one of four named DuPont chemical engineers for work at the Met Lab's technical development program.

DuPont's Harrington and Chilton offered the services of Charles M. Cooper who had been Director of MIT's School of Chemical Engineering before joining DuPont in 1935. Cooper reported to Chicago on August 3, 1942, remaining there for the duration of war.[1, a]

Several weeks later, unknown to DuPont, Groves was considering the possible methods for managing the plutonium-program.

"Next, I had to decide whether to place the various responsibilities for the engineering, the construction and the operation with one firm or with several," he wrote. "To me, a single firm carrying the threefold responsibility seemed by all means preferable. For one thing, it would lessen the problems of co-ordination that would fall into my lap, which by that time was becoming rather crowded with major problems. After I had studied all the possibilities, I concluded that only one firm was capable of handling all three phases of the job. That firm was DuPont."

"A further attraction of DuPont," he continued, "was that for several years I had been working most successfully with its engineers on the Army construction program."

And finally, he remarked, "A number of us had come to realize by that time that the operation of the plutonium plant would be extremely difficult, and that it would require highly skilled technical management, thoroughly experienced in large operations. A thorough knowledge of chemistry and chemical engineering was important, and this DuPont had."[3, b]

Groves had understood that DuPont's prime qualification was its basic nature, a topic explored next in greater detail.

---

[a] Subsequent to his arrival, Cooper's role became major indeed. He became a member of the Laboratory Council and Supervisor of the Technical Division. The latter position placed him in charge of: engineering of the separation-process flow sheets; metallurgy, and the six metallurgical subcontractors; and control and instrumentation. [2]

[b] Implementing his decision to select DuPont, Groves met the Board of Directors on September 28, 1942 to request DuPont's assistance in design and procurement for the separation plant, then planned for Chicago. In subsequent meetings on October 10 and October 30 Groves overcame DuPont's understandable reluctance to undertake the design and construction of the complete plutonium plant, including the piles. On November 12, the Executive Committee agreed to take on the complete plutonium project, saying that in the light of the national emergency, they could not in good conscience refuse.[4] (See Matthias' Chapter 10 description of the final meeting.)

The Nature of DuPont

An understanding of DuPont as it existed in 1942 is fundamental to an understanding of what happened at Hanford, a concept starkly illuminated throughout the book under such topics as Engineering, Procurement, Costs, et al.

DuPont's primary characteristics at that time, as they related to Hanford were the qualities of their departments of research, engineering, and construction.

Research

Walter S. Carpenter remarked in 1945 that DuPont was able to carry out its wartime assignments largely because of the research previously carried out in peacetime.[5]

DuPont had strongly emphasized research since the inception of the modern DuPont company in 1902. Seventy percent of the Executive Committee then had engineering or technical training, including Pierre DuPont, an MIT chemistry graduate.[6] In the next forty years leading up to Hanford the company came to regard research as the principal basis for new products and expanding income.[7] In 1941, the year prior to Hanford, they spent 21.6% of their earnings to support 1341 research personnel.[8]

DuPont's outstanding chemist, Wallace Carothers, led the nylon research program, was considered the best organic chemist in the U.S. and was a member of the National Academy of Sciences.[9]

Engineering – *(Pre Hanford)*

At the urging of Charles Stine, Director of the Chemical Department, the Executive Committee established a pure scientific research group in 1927. The chemical-engineering branch of the new group, under T.H. Chilton, then did fundamental research in: the unit operations of process systems; properties of substances; the analysis of systems on the basis of chemical properties; heat transfer; and mass transfer. Allan Colburn, a member of the group, won the first Walker award of the AIChE for his analog of heat transfer to fluid flow.[10]

Raymond P. Genereaux, later a notable engineer on the Hanford design, was assistant manager under Chilton, and later manager. Genereaux remarked that "We had a golden opportunity, really, because no one else had approached the subject as we did. One of our main objectives was to produce results that could be used by practicing engineers in the company, and by our publications (this information) spread outside, which helped in recruiting good people and made a contribution by the DuPont Company to science and engineering as well."[11]

DuPont's John Perry originated the Chemical Engineers' handbook in 1934. Thirty percent of the authors of the second edition, the edition current at the start of Hanford, were DuPont chemists and engineers. George Graves[a] and Hood Worthington were two of the authors, and were also prominent in the Technical division, headed by Crawford Greenwalt,[b] of DuPont's Explosives-TNX, the ad hoc department responsible for the management, technical analysis, and operation of the Hanford project.

---

[a] Graves, in 1936, had solved the fiber-breaking problem in nylon melt spinning by pressurizing the spinning pumps, thereby preventing bubble formation.[12]

[b] Greenwalt knew nothing of nuclear physics in 1942, but had acquired enough physics by the time of the Xenon-poisoning incident at B Reactor independently to deduce the cause.[13] Seaborg was "particularly well impressed by Greenwalt's grasp and understanding" of nuclear physics after only three weeks on the job.[14] Greenwalt's historic notes of Dec. 2, 1942 are shown in Fig. 5.1.

EXCERPT FROM C.H. GREENEWALT'S
NOTEBOOK #3

On Wednesday afternoon, 12/2/42, Compton took me over to West Stands to see the crucial experiment on Pile #1. When we got there the control rod had been pulled out to within 3 inches of the point where k would be 1.0. The rod had been pulled out about 12 inches to reach this point. The resultant effects were being observed (1) by counting the neutrons as recorded on an indium strip inside the pile (see previous notes) and (2) on a recorder connected to an ionization chamber placed about 24 inches from the pile wall. The pile itself was encased in a balloon cloth envelope. The neutron counter was not a good index of what was going on since the number striking the indium strip was near and above the number which could be counted with accuracy. Hence the best index was the recorder attached to the ionization chamber. This had two ranges - one about twenty times as sensitive as the other. Fermi was cool as a cucumber - much more so than his associates who were excited or a bit scared.

The recorder curve when I arrived was showing the effect of the last move (12 inches to within 3 inches of k (kappa) - 1) It was working on the sensitive scale. The curve was about as follows:

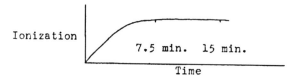

The recorder was then moved to the insensitive scale and the control rod pulled out 6 inches to a point estimated to be 3 inches over the position for k - 1. The curve drawn was very nearly a straight line and it was difficult to tell whether the ionization was increasing or falling off. (The count was too fast for the neutron counter to be of use.) The curve was about as follows:

At this point, Fermi pushed one of the other control rods all the way in and pulled the principal control rod out 12 inches. The ionization fell to 0 immediately and the clatter of the neutron counter practically stopped. The other control rod was then pulled all the way out and off we went. This was at about 3:25 p.m. Chicago time. The neutron count quickly showed that the rate of emission was increasing with time and that we really were "over the top", k greater than 1.0.

The ionization recorder soon showed the same thing (insensitive scale) and kept increasing by leaps and bounds. Affairs were allowed to go on until an automatic release was cut in which shot in a control rod and stopped the reaction. This it did suddenly and completely. So there we were - for the first time - chain neutron reaction had been demonstrated. Fermi was still calm but happy. There was applause and he was given a bottle of wine (in a paper bag). The curve (about 10 minutes over all on the recorder) was as follows:

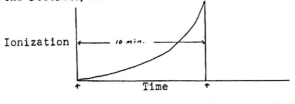

Control Rod pulled out

Automatic release functioned

Compton said that results obtained in this experiment indicated that the amount of U in subsequent piles might be cut in half, i.e. 60 - 100 tons. Aside from demonstration that the reaction proceeds the most important point is that ease of control (at room temperature) with minutes to spare has been clearly demonstrated.

The neutron concentration in the room during the "power flash" rose to well above the tolerable limit and the gamma radiation to just about the tolerable limit. (For human exposure 24 hours per day)

The overall result was much better than expected. It was for me a thrilling experience.

Figure 5.1

CRAWFORD GREENWALT'S NOTES
of
THE FIRST NUCLEAR CHAIN REACTION IN HISTORY
December 2, 1942

Crawford Greenwalt was one of DuPont's representatives on the Lewis Committee, which was making its second inspection trip to the Met Lab on December 2, 1942. This was the same day that Enrico Fermi had Pile No. 1 ready for his first try at a chain reaction.

Arthur Compton, the Director of the Met Lab, wanted a Lewis Committee member to witness the event, and picked Greenwalt because he was the youngest and would therefore remember it the longest.

These typed notes of Greenwalt's are from his laboratory files at the Hagley Manuscript Library, Wilmington, Delaware.

Reference: End Note 15

Graves wrote on organic chemistry, Worthington on high-pressure technique, and Genereaux on fluid flow. Stine was on the sponsoring board for the Handbook and Chilton and two other DuPont men were on the editorial board.16

Genereaux, Colburn, and Chilton, while at Dupont, taught graduate courses in chemical engineering at Columbia. Genereaux was asked to head Columbia's and Pennsylvania's chemical engineering departments, but decided to stay at DuPont. According to Genereaux, DuPont would have lost their entire research group if everyone in it had accepted the university offers made to them.17

Roger Williams, manager of the TNX department during the Hanford project, at the end of WW I had created the Ammonia department's technical organization, and managed the process development of long-chain alcohols, urea, acrylic resins, and high pressure systems technology at DuPont's Belle, West Virginia plant.18

The 1948 edition of "Who's Who In Engineering" included Chilton, E. G. Ackart, DuPont's Chief Engineer during Hanford, and T.C. Gary, manager of the Design Division.19

Construction – (Pre Hanford)

In the years prior to Hanford, DuPont had been engaged in a very large war-plant construction and operation program – smokeless powder plants and other facilities – for both the U.S. and Britain. This program, worth $750 million, required 182,000 people on 18 projects in 14 states and the spending of $1 million per day for labor and materials.

This work was completed on very short schedules, typically: 4.5 months for a 25 mw power plant; seven months for two powder plants; and 12 months for an ammonia plant.20

By the end of WWII DuPont had built 65% of total U.S. Ordinance Department powder production.21

This ordinance Department work done under the direction of the Corps was a principal basis for General Groves' opinion that DuPont was the primary candidate out of all U.S. contractors for the unprecedented type of process construction envisioned for the plutonium plant. Matthias remarked, "They had done such good work for him – amazing work!", and "DuPont at that time really had a program on construction, very thorough, and they cut no corners on any job!" 22

DuPont Veterans Comments On Engineering And Construction

The DuPont professionals of that time who were interviewed for this report remarked that DuPont had decades of experience in engineering corrosive and toxic processes and, at Belle, West Virginia, systems at 1000 atmospheres. Simon, Genereaux, and Tepe pointed out that DuPont exercised the same degree of engineering care and caution at Belle that they did with the also-unprecedented Hanford design. They considered that DuPont had a very strong engineering organization and that one of their strengths was in having research and plant-operating people working with design. DuPont, said Genereaux, made a point of continually updating themselves on new process machinery, and because of the corrosive fluids with which they worked they constantly studied new materials.

They were unanimous in emphasizing another major DuPont characteristic: everyone in the company knew each other very well – their capabilities and past performances – and because of this familiarity they could work with, and cooperate with each other with confidence and without having to protect themselves in writing. Simon stated, "We had excellent communications, and we could say what was on top of our minds without being careful of everything we said, and because of this, everything moved fast."

Tepe said, "It was a pleasure working there because when a man was assigned responsibility, you could count on his discharging it." He also observed, "I do believe that DuPont had a capability from research through construction that was equal to anybody in the world - I'd say the best in the world."

Simon remarked, "DuPont didn't use any new management methods at Hanford. They just followed the procedures they had used over the years." Remarking on DuPont's high morale, he said, "We had the saying at DuPont, 'Rome wasn't built in a day, but then, DuPont didn't have that job.' "

It was Stanton's observation that DuPont's construction engineers and craft superintendents had worked together for many years and, as a result, had a great deal of cameraderie. "We knew we had a great construction department and we were proud of it," he said. He also remarked, "DuPont had a tremendous engineering organization. We had internationally renowned experts in every phase of engineering." He further stated, "DuPont engineering had every phase of engineering in house – heat transfer, materials engineering – everything."[23]

From the above recitals it is apparent that in 1942, the firm that was to manage Hanford comprised a closely coordinated, long-standing, synergistic combination of pure science, research, engineering, plant operation, and construction, with a proven, four-decade record in process-plant work

The details of DuPont's management of the Hanford project are discussed next, beginning with the firm's general organization at Wilmington – then progressing through engineering, procurement, and Hanford field management.

• General Organization At Wilmington •

DuPont in 1942 comprised two principal branches reporting to the President: manufacturing, with 10 departments,[24] one of which was Explosives; and auxiliaries, with seven departments,[25] one of which was Engineering.

In the Fall of 1942 a new manufacturing department was added, Explosives-TNX, to manage DuPont's plutonium work, to coordinate Met Lab, Wilmington, and Hanford activities, to provide technical analyses to Design and Construction, and to manage Hanford operations upon the start of production. Figure 5.2 displays a simplified organization chart of this new department.

Roger Williams was appointed as director of TNX. Although he bore the title of Assistant General Manager, TNX was independent – an eleventh manufacturing department. E.B. Yancey, the General Manager of the Explosives Department was available for consultation however.

Within Engineering, the Design Division was grouped in accordance with DuPont's manufacturing departments, e.g. Ammonia, Organic Chemicals, etc., plus a core of discipline groups. A new design group for the plutonium project was added – the Explosives – TNX group.

DuPont's Construction division, rather than reporting to the President, as is the case for many AE-constructors, reported to the company's Chief Engineer, as did the Wilmington Shops and other divisions concerned with plant design, maintenance, and operations. A Hanford construction group was added in 1942 to the existing War Construction subdivision of the Construction Division.

The relationship – Engineering in charge of Construction – extended down to every job site, including Hanford, as it had done for many years.[27] The job-site implementation of this relationship is subsequently explained in detail in the section on Hanford Field Management. Figure 5.3 is a simplified organization chart of DuPont's Engineering department.

Figure 5.2

EXPLOSIVES DEPARTMENT TNX
SIMPLIFIED ORGANIZATION CHART
June 10, 1944

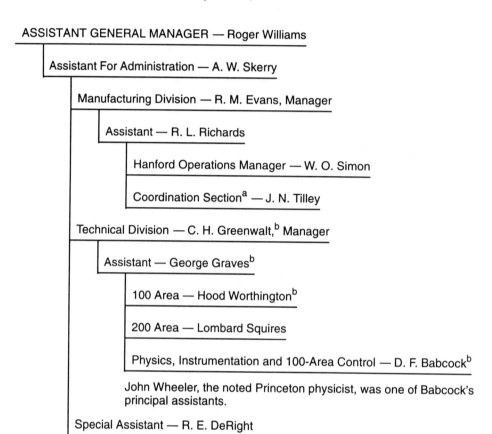

- ASSISTANT GENERAL MANAGER — Roger Williams
  - Assistant For Administration — A. W. Skerry
    - Manufacturing Division — R. M. Evans, Manager
      - Assistant — R. L. Richards
        - Hanford Operations Manager — W. O. Simon
        - Coordination Section[a] — J. N. Tilley
    - Technical Division — C. H. Greenwalt,[b] Manager
      - Assistant — George Graves[b]
        - 100 Area — Hood Worthington[b]
        - 200 Area — Lombard Squires
        - Physics, Instrumentation and 100-Area Control — D. F. Babcock[b]

      John Wheeler, the noted Princeton physicist, was one of Babcock's principal assistants.
    - Special Assistant — R. E. DeRight

      DeRight maintained a log of plutonium-project events and kept a chronology, a portion of which is listed in the References, Appendix E.

---

[a] Technical Division directions were formally transmitted to Hanford Operations and to the TNX Design group by this Coordination Section, a relationship identical to that prevailing in the other ten manufacturing departments. Informal transmittal however, was continuous in frequent meetings.

[b] Key men in the development of nylon.

Reference: End Note 26

Figure 5.3

ENGINEERING DEPARTMENT
SIMPLIFIED ORGANIZATION CHART
June 10, 1944

- CHIEF ENGINEER — E. G. ACKART
  - Assistant Chief Engineer — G. M. Read
    - Design Division — T. C. Gary, Manager
      - Supervising Engineer — F. W. Pardee, Jr.
        - Explosives — TNX design group — H. T. Daniels, manager
          - Assistant Design Project Managers
            - R. P. Genereaux — Separation Areas
            - J. A. Burns — Reactor Areas
            - L. H. Haupt — Other Areas
      - Eight other industrial-department design groups
      - Discipline groups
    - Construction Divison — M. F. Wood, general manager
      - (Managed by G.M. Read in 1943 until he was promoted to Assistant Chief Engineer)
      - War Construction — F. H. Mackie, Manager
        - HEW Field project manager — G. P. Church
        - Other war projects
    - Technical Division — T. H. Chilton, Director
    - Wilmington Shops Divison — W. R. Heald, Manager
    - Other groups: estimating, industrial engineering, mechanical and process improvement, plant maintenance, electric power, water and waste treatment, chemical engineering, metallurgy, machine development, etc.

Reference: End Note 28

DuPont And Turnkey Management

DuPont had routinely managed its engineering and construction forces by the turnkey method at least since 1914 when they expanded their powder-plant capacity by 1300% to meet the European Allies' desperate need for smokeless powder in World War I. America's entry into the war in 1917 required that DuPont engineer and build, again by the turnkey method, still more powder capacity – the enormous Nitro, West Virginia and Nashville ("Old Hickory") plants. All of these plants were simultaneously designed by DuPont's Design Division while their Construction Division was building them. Management of these projects was by DuPont's Explosives Department, the manager of which functioned as the turnkey Project Manager, who coordinated design and construction while also serving as the link to DuPont's client, the War Department. (The story of these fantastic efforts in spite of extreme administrative foot dragging by the War Department is a fascinating account of proficient turnkey management.)[29]

This same relationship was in effect during the twenties and thirties in the design and construction of their own process plants. DuPont's cognizant manufacturing department – Ammonia, Fibers, etc. – managed the activities of the Engineering Department's Design and Construction Divisions.

For Hanford, Roger Williams functioned as the Project Manager for planning and coordination of the turnkey operation. TNX and Williams were also the link with the client – the MED. The MED offices at Washington, D.C., Wilmington, and Hanford contacted TNX for administrative matters, to transmit instructions, and for information.

The relationship between the ad hoc TNX and the long-standing Engineering Department was the type of turnkey organization identified by Barrie and Paulson as a "task force within a matrix organization," in which such key figures as Ray Genereaux and Gil Church reported to both the task force and to the matrix. Genereaux, for example, received conceptual input from Squires of the task force, while at the same time having available, through Daniels, et.al., all of the resources of the Engineering Department matrix.

As will be seen in subsequent sections on Engineering, Procurement, Wilmington Shops, and Hanford Field Management, turnkey management by an engineer-constructor was the only method that had a chance of timely completion of Hanford under the demanding constraints imposed by the necessity of parallel science, design, procurement, and construction.

Other types of management than turnkey would have been either inapplicable or too time consuming, as follows:

The Owner-Builder type was not applicable because the owner – the Corps' MED – had neither a process engineering arm nor a process contracting arm; that is why they hired DuPont.

Professional Construction Management features a three-party team of manager, separate designer, and separate contractor. The addition of an outside professional construction Manager to the Hanford-Wilmington organization would have served no purpose and would have delayed the project because of the need for coordination between the outside manager and DuPont. It is obvious from Figures 5.2, and 5.3 that such an outside agency would also have been redundant because DuPont had their own internal management.

Delay would also have been experienced in the Traditional Method of Owner plus separate contractor and separate engineer. The time lapses in coordination between these three separate entities, plus the required lead time for design would have caused far too much delay to meet Hanford's demanding schedule.

Raymond P. Genereaux

Assistant Design Project Manager for the separation plant. Taken about 1955 in his office in DuPont's new Louviers Building near Newark, Del.

RAYMOND P. GENEREAUX

Left: **Gilbert Church**, Field Project Manager at HEW

Center: **Granville M. Read.** For HEW he was Manager of the Construction Division, then Assistant Chief Engineer. At the time of this photo, January, 1948, he was DuPont's Chief Engineer.

Right: **Russell C. Stanton**, Division Engineer in charge of the 101 Building and the three 105 Reactor buildings.

Taken outside the Desert Inn at Richland. DuPont met with General Electric on providing know-how on construction of new reactors.

RUSSELL C. STANTON

Lombard Squires

Head of separation-process coordination in the Technical Division of TNX. Taken in the mid Fifties.

LAWRENCE BERKELEY LABORATORY/BATTELLE PRESS/ THE PLUTONIUM STORY

Crawford Greenwalt

Manager of the Technical Division of TNX. Later, President of DuPont.

LAWRENCE BERKELEY LABORATORY/BATTELLE PRESS/ THE PLUTONIUM STORY

The application of DuPont's long-standing turnkey organization to the plutonium project is discussed next, beginning with Engineering.

• Engineering •

Hanford's engineering-development sequence was as follows: development of the scientific and conceptual bases and flow sheets by the Met Lab; TNX transmittal of these data to DuPont's Design Division for physical engineering and design; TNX monitoring of design; TNX transmittal of changes in requirements; and coordination by TNX of the Met Lab schedule with DuPont's schedules for construction and operation of the Hanford plant.

Engineering Perspective

It was Walter Carpenter's opinion, referring to HEW in 1945[30] that "the contributions of the Engineering Department were outstanding and the design and construction problems met and solved by it were the greatest the company ever encountered."

In making this comparison, Carpenter said, in effect that Hanford was more difficult than DuPont's engineering and construction of the mammoth Old Hickory powder plant in WW I, their 10-year nylon development, their 15-year neoprene development, and their Belle plant designed to operate at 15,000 psi.

"Outstanding," when applied to the Hanford project was not an exaggeration. DuPont's engineering resulted in a nearly flawless startup on a very short construction schedule despite the exotic, first-of-a-kind nature of the process and its components, descriptions of which were provided in Chapter 2, and which are further detailed in this chapter.

A break-in period of some length of time – months or even years – is the usual experience with commercial process plants. Richard Doan, Chief Administrative Officer at the Met Lab, remarked in a planning session there in November, 1942 that "experience in the Phillips Petroleum Company indicates that at least two years is necessary to get into production after the fundamental research has been done, and this on processes much simpler and better known than ours."[31]

How did Hanford avoid such long delays? One fundamental reason was the nature of DuPont, as previously described – the firm thoroughly understood process engineering. DuPont's engineering practices, subsequently described, were another significant part of the explanation.

But beyond those reasons were two basic planning decisions jointly made by DuPont, the Met Lab and the Corps.

Basic Planning Decisions

The first basic decision, in the Fall of 1942, *was to gamble on an immediate start of design and construction*, despite the fact that reactor design had not even reached the pre-conceptual stage and that the separation process might be any one of twelve concepts, all of which were incompletely developed.[32]

Furthermore, no one had so much as seen the product, and wouldn't for another 15 months, and then only in microgram amounts.[33]

The second basic decision was to design and build the plant and to develop its unprecedented components and processes *in parallel with each other, with the development of the supporting science, and with the design and operation of the semiworks*. There was, of course, no advance assurance under this parallel program that each component and process would not only work as planned, but would be ready in time to function properly with the other components and processes.

This paralleling of scientific research and semiworks operation with conceptual design, final design, procurement, and construction is the ultimate extreme of what is today termed "phased construction."[34] What was being done was to begin final physical design before either the laboratory semiworks or the pilot plant had even been started, with only incomplete laboratory findings as the basis for final design![a] Neither DuPont nor any other turnkey constructor would have deliberately adopted this procedure by choice. According to Genereaux, DuPont's normal, peacetime commercial practice was to hold off on construction until final design had reached 60% completion.[35] It is obvious therefore, as observed previously, that those responsible for the project had been forced into desperation measures.

Simon, Hanford's Operations Manager, remarked that the question of pulling all of these parallel efforts together for a successful startup "was a constant worry."[36] Tepe, on the other hand, and others at the research and process-design level held a more sanguine view. "I think the scientists fully recognized (these risks) but the attitude which I always observed was optimism and confidence and faith that they could solve the problems, and it would work. And that applied not just to the nuclear project, but to the commercial work. Commercial work involving new processes and new products took similar risks."[37] In other words, the hands-on scientists and engineers who were in the position of doing something concrete with the problems believed they could be solved.

Genereaux made a related remark when I asked him if the appalling difficulties of the new separation process, particularly with little data available, caused him concern. "I never took that attitude about any job," he replied. "It was just a job to be done."[39]

Developing further the problems of the parallel approach, I display some of the more prominent parallel tasks in Table 5.1. Also, Figure 5.4 shows the chronology of some prominent development events in parallel with the drawing and manpower schedules. Some very short lead times are seen in that figure between last minute developments and plant startup. (Table 5.1 was taken from DuPont A and from Seaborg A. Figure 5.4 was from the same, from the Design Status graphs of DuPont B, and from Matthias B.)

Some of these parallel activities were completed early. For instance, the final decision for a cubical, graphite-moderated, light-water-cooled reactor with uranium slugs in aluminum tubes was made on February 16, 1943, only five months into design, and a month before start of construction.

Other key activities and components however, were not resolved until quite late, the most difficult being the slug-canning method, which squeaked in only a month before it was time to load the B Reactor. Simon commented, "They just about barely had enough cans to charge the first pile when it was ready. It was a close call."[40] Simon's account in his interview of how the slug-canning was solved is of interest because it differs from some of the other accounts, the "midnight inspiration" explanation, for example. His recital is found in Appendix H-2, 10.WS through 12.WS

The fuel-tube manufacturing questions were not resolved until quite late – December 14, 1943.[41] Final revisions to the cell part of the process were not made until June 20, 1944, over 14 months after completion of final design of the cells, and two months after equipment installation had begun in the canyon at 200 West.[42]

---

[a] Because of this undesirable but necessary timing, the Clinton separation semiworks was of very little use in the design of the Hanford separation plant. In some cases in fact, Clinton used the Hanford design experience in the completion of Clinton's separation plant.[38] As noted below, Clinton's usefulness to Hanford's separation design was in the results of Clinton's operating experience, plus of course, their contribution of metallic plutonium to the Met Lab studies.

Table 5.1

PROMINENT QUESTIONS PARALLELING DESIGN AND CONSTRUCTION

In September, 1942, when it was decided to proceed immediately with design, procurement, and construction, the following were in a state of flux, or had not even been started.

FOR PILE

1. Continual re-calculation of multiplication factor, k, as the result of changing decisions on geometry and materials.
2. Reactor type and geometry possibilities: Uranium rods, water pipes, and graphite? Uranium lumps, or continuous fuel channel? Metal pipes, with graphite and uranium shot as coolant? Spherical, cylindrical, or cubical pile? Graphite or heavy-water moderator? Lattice spacing? Control-rod type?
3. Intermittent or continuous operation?
4. Uranium oxide or metal for fuel?
5. Air, CO_2, water, helium, or bismuth coolant?
6. Uranium metallurgy and machinability
7. Design, construction, and operation of the Clinton pile to obtain design data for Hanford.
8. Design, construction and operation of the Clinton SMX.

   This unit didn't begin operation until August, 1943 – then ran until May 1944. During this period, the 105 Area drawings went from 6% complete to 87%, and by May of 1944, B Reactor construction was 70% complete.

FOR SEPARATION PLANT

Note that in September, 1942, separation-cell physical design was to begin almost immediately.

1. Which of the 12 possible separation methods shall be used?
2. Fission-products chemistry, and the nature and amounts of fissions products to be expected?
3. The questions of purification of uranium required because of alpha reaction on light element impurities.
4. Design, construction, and operation of the Clinton separation plant to obtain data for selection of Hanford cell equipment, and to produce plutonium metal for Met Lab use.

FOR PILE AND SEPARATION PLANT

1. Radiation effects and health physics.
2. Radiation testing of pile materials and of extraction chemicals.
3. High-temperature testing of pile materials.

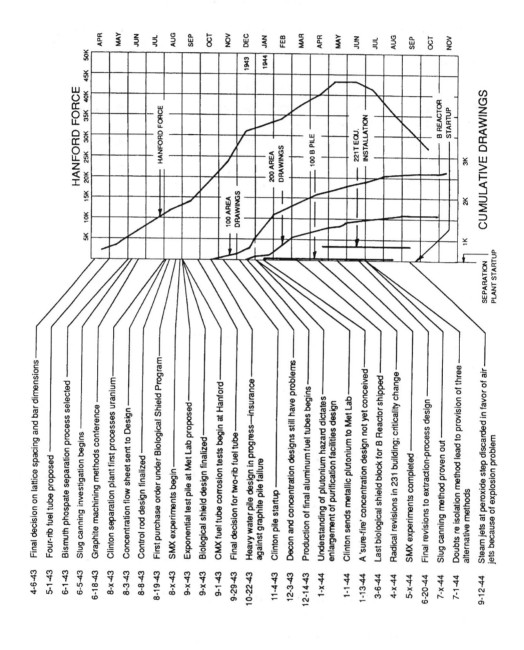

Figure 5.4

**PARALLEL DEVELOPMENT EVENTS
vs. DRAWINGS, MANPOWER, AND CONSTRUCTION PROGRESS**

Changes of this type however, had long been anticipated by Wilmington design, as described below.

Clinton discovered an explosion problem in the use of steam jets for peroxide transfer in the concentration step, requiring a September 12, 1944 change to air jets, only three months before separation-plant start up.[43]

These two basic planning decisions were, of course, virtues born of necessity. Had the MED and its contractors played safe and, instead of parallel activities, gone to the rational, semisequential route – science first, followed by conceptual and preliminary design, semiworks, final design, procurement, and construction – Hanford completion would have been far too late to affect the war. How much too late is an interesting topic which can be calculated to a fair degree of accuracy, as shown in Appendix G and as discussed next.

Parallel And Rational Schedules

I investigated the question of schedule stretchout, assuming that the plutonium project, instead of employing the parallel method, had been planned according to rational design/construct practices. The results are shown in the three bar charts of Figure 5.5, which sumarize three types of planning:

- **Parallel Wartime.** What actually happened, with parallel activities and almost everyone working overtime.
- **Rational Wartime.** What would have happened with the same amount of overtime as was actually used, but if rational schedules of activities had replaced parallel activities.
- **Rational Peacetime.** What would have happened with no overtime, in peacetime, and with rational schedules. I modified the arithmetic to allow for some minimal delays caused by Congressional oversight at milestones. Probable peacetime labor disputes, which I did not attempt to forecast, would have lengthened the peacetime schedule further than shown on the bar chart.

Under the rational, wartime schedule the first plutonium bomb would have been delayed until May, 1948, two and one half years after the scheduled Kyushu invasion. The war may well have been over by this time, but even if it were still in progress, the presence of American troops would have probably prevented the use of the first plutonium bomb.

Finally, the Rational Peacetime schedule is of interest because it approximately validates today's Defense Acquisition cycle. Today's weapons require 15 years after appropriation of final-development funding before the weapon is deployed. This compares with the 12 years in Figure 5.5 for all activities subsequent to completion of the basic science.

DuPont's Engineering Practices

DuPont's underlying approach was that of detailed and painstaking care and caution. Genereaux observed, "We were successful in not having any failures because we paid total and constant attention to detail, and because we did not spare good design and detail development in order to save time." In this connection Simon remarked, "Frank Matthias' engineers were stunned by DuPont's cautious and careful approach to Hanford's design." He said that on one occasion Matthias tried to get DuPont to quit turning square corners by saying, "Henry Kaiser built a ship in four and a half days," but DuPont still plowed ahead with their pre-set method.[44] (Given Hanford's recorded schedule and costs however, DuPont must have turned their square corners at a high velocity!)

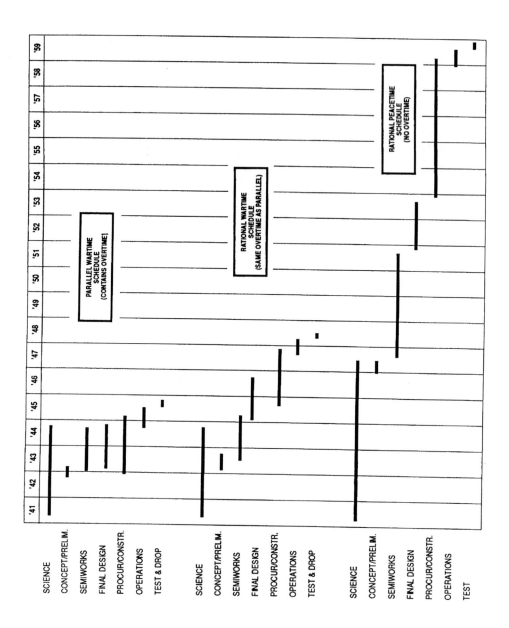

Figure 5.5

PARALLEL AND RATIONAL SCHEDULES
(See Appendix G for rationale & sources)

Everything was designed and drawn. For example, piping smaller than two and one half inches diameter was dimensionally located on the drawings rather than the frequent practice of leaving it to pipefitters' discretion and standard practice. The design was operationally realistic because, in common with DuPont's commercial practice, operations people worked with Engineering during design. Elaboration was avoided, as it was in the design of their commercial plants. "We tried to have everything as simple as possible," said Genereaux. "The least you have in design the better off you are. It's a matter of probability. The more pieces, the greater probability of something going wrong."

Simon agreed with Genereaux that the painstaking approach used at Hanford was DuPont's normal practice for difficult work. They took the same approach for unusually demanding commercial work, such as the ultra-high-pressure Belle chemical plant.[45]

Checking

Everything was checked. "Drawing checking was by a separate group," Genereaux remarked. "There was no such thing as spot checking. The checkers ran through all the design steps that the original designers did, so that 100% of the design effort and of the drawings was checked. DuPont's direct experience with other engineering firms was that many didn't understand true, detailed checking."

George Graves commented on the checking within TNX. "I learned one thing from Greenie (Crawford Greenwalt, his boss) and that was how to run a plant on paper. I guess he had learned that as a chemical engineer – to take the designs and the plan and follow it through, valve by valve, pipe by pipe, to see that the product could get from where it was supposed to be to where it was supposed to go. I remember we did that for the separation plant, valve by valve."

W. K. MacCready was a Division Engineer in the construction of the separation plants, and in describing the startups there said, "I had been involved with startup at about 15 plants, and none of those plants, which were relatively small and simple by comparison, started up as easily and in as trouble-free a fashion as these out here did. I think that was . . . because they were impressed with the significance of the fact they were going a long ways on very little information. So everything was most carefully checked. Every step was double checked. As a consequence there was essentially nothing that had not been accounted for."[46]

I made a point of asking Genereaux and Squires about startup difficulties that might have occurred beyond those reported in the literature. "Did you really have as few difficulties as MacCready reported?" I asked. Genereaux replied, " I didn't hear of any problems, and I surely would have been the first to be told." All Squires could think of was the discovery prior to hot startup that the specified Teflon gaskets had insufficient shear strength and had to be replaced with blue asbestos. Additionally, Gerber reported that the centrifuge hydraulic hoses, originally supplied with brass connectors had to be replaced with steel-connected hoses.[47]

For perspective, note that MacCready, Genereaux, Squires, and Gerber were reporting this essentially perfect startup in $775 million (1994 dollars) worth of separation plant![a]

---

[a] With the reactors however, a number of design and operational problems were discovered almost immediately after startup, many of which could have been worked out had there been time first to build a water-cooled pilot plant. These problems, described in Gerber A, pp. 9-20, included the slug discharge facilities, fuel-tube cooling-water flow, flux anomalies, dummy-slug materials, etc.

Speed Of Design

"It was very exciting. It was going so fast that we talked it out over the drafting boards. We wrote very few memos", said Genereaux. "I told my engineers never to work with more than three significant figures; two were usually sufficient. With that criterion, *we designed the separation plant with six-inch slide rules.*" (Emphasis added)

Simon commented on the velocity of Genereaux's group. "There was a lot of speed in their work because everyone knew each other well enough that it was handled verbally – everything wasn't written down. In fact, verbally it moved so fast that it worried some people. 'How do we know what's going on?'"

"We maintained a good pace," said Genereaux. "My standard was 16 days to do a drawing for commercial work. We couldn't always do that at Hanford because of the very large amount of development work we had to do at the Wilmington Shops and at the vendors." Table 5.2, compiled from DuPont A, displays a list of some of the more prominent design development items that were charged to the Design-Division budget. Schedule delays and added design costs obviously resulted from the items of this list.

"I knew General Groves quite well during the project, and he was a great help. One time he asked if there was anything he could do to help me, and I said, 'Yes, keep everyone off my shoulders.' By that I meant to keep all the ribbon clerks in the project from slowing down approval of my drawings. So he spoke to a number of people, and my drawings were expedited through Corps approval in their Wilmington office."[48]

Drawing List And Schedule

"We started out with a drawing list, then added drawings as they became necessary," Genereaux said. "Then we finished with a 'negative' list which was a list of all the items that were yet to be completed – similar to a construction punch list." Wilmington's design and drawing schedule led the field, not vice-versa. We knew the proper construction sequence – then cranked out the drawings as fast as possible in that sequence. That kept the project moving. We issued as few preliminary drawings as possible. When the field asked for them we issued them, of course, except for the separation cells – we never gave out preliminary drawings for them, or very few of them."

There were some notable preliminary drawings however, at the start of construction – the dimensionless, general excavation drawings with which the 100 B Area site excavation was started. Wilmington filled in the dimensions as fast as they could by teletyping the numbers to Hanford.[49]

Design Security

A good, summary statement of Hanford security was that in the DuPont/Corps Technical Record of Negotiations;

"In designing and laying out the site, and in designing and constructing the plant, it has been necessary to take into account the possibility of sabotage and other enemy action. Stringent security measures have governed and will govern all research, design, procurement, construction and operation procedure, in order that the utmost secrecy may be maintained. Relatively few individuals engaged in the work have full knowledge regarding it. Other personnel is divided into groups, each possessing only such information as is necessary for its own use in connection with its own part of the work."

Table 5.2

HANFORD DESIGN-DEVELOPMENT PROGRAMS CHARGED TO THE WILMINGTON DESIGN ENGINEERING OFFICE

- 70 special fabrication problems on the shield-block program (See subsequent section on "Procurement")

- Over 40 Wilmington Shops developmental and test items. Furthermore, many of these "items" were programs that included many tests and other shop tasks under the single item

- Wilmington Shops separation-cell mockup testing of Design Engineering concepts; backcharge to Design for this effort

- Backcharge to Design for construction cost of the cell mockup

- Vendor coordination effort required by cell equipment precision to hundredths of an inch

- The 12-month development program for uranium slug extrusion and machining

- Part of the 18-month development of the slug-canning method

- 12 months of consultation with three vendors on fuel-tube development, plus designs of alternative tubes

- Consultations with graphite manufacturers

- Consultation on graphite-block design and on block machining methods

- Study and testing of anti-aircraft gun drives as control rod drives. Negative results; study discarded

- 3-month study of electrical control-rod drives and rod materials

- Year-long development of final control-rod drives and rod materials

- Construction cost of the Hanford CMX water-corrosion study

In implementing this policy, every worker in every category was investigated by the FBI as to nationality, birth, character, and social and political activities.[50]

Genereaux recalls that "We had secure drafting rooms and you had to have a security pass to get into them. One or two of us had to go around at night to see that everything was locked up. We had no secure phones. Instead, we used very obscure terminology when we talked on the phone. We carried these terms in our heads after having first discussed them with those who would be using them. We never wrote them down. And we operated on a strict, need to know principle. We learned to work around all the problems that security caused us."

As an example, he cited the unusual, security-related features on some purchased equipment and their related drawings. "I told the vendors what we wanted, but not why," said Genereaux. He further stated that security did not adversely affect the speed or quality of work, efficiency, or morale. In other words, security did not have a major effect on the project.[51]

Separation Plant Physical Design – *Basic Engineering*

As stated earlier, a principal means of expediting the project was to pursue science, design, and construction in parallel. In October, 1942 when DuPont first began designing the separation plant, the Met Lab was investigating 12 separation concepts (as described subsequently in Chapter 6).

As these investigations proceeded, some process decisions had been made by January of 1943: it would be a liquid process; the process would be one of the precipitation methods; it would be at room temperature; flows would not be excessive; a principal piece of equipment would be a Bird centrifuge; and all precipitates would be re-dissolved before leaving the cells, so that inter-cell streams would be clear liquid. The dimensions of the cells and covers were fixed. The cover thickness was fixed by the radiation shielding requirement. The cover was divided into segments to match the capacity of the canyon crane, By this time the structural design of the cells was firm.

Although the Met Lab and TNX had, by early spring of 1943, furnished the Design Division with updated flow sheets for the alternative processes, Ray Genereaux had had incomplete design criteria for vessels and for piping connections. The final decision for bismuth phosphate would not be made until June of 1943 and, as it developed, the Met Lab would make at least six major process changes in the cell equipment in the 19 months prior to August of 1944 (See Figure 5.4).

The procurement and construction schedules however dictated that Genereaux complete his final design of the cells by the end of March, 1943.

He therefore made some assumptions as to reasonable capacities for vessels and other equipment. He then assumed reasonable fill and discharge times, thus sizing the pipes. "It was a judgemental thing," said Genereaux, "We were experienced in dealing with chemical processing equipment, and we estimated our piping, utility, and layout needs based on that experience."[a]

He kept the cell arrangement flexible and decided on an ample number of cells, "We designed a 'kitchen' in which we could cook anything when they finally decided what the process was to be. We had standard, remotely installable connections for electricity, water, gases, lubrication, and process fluids – 42 connections per cell – so they could later design in anything – mixers, vessels, centrifuges – anything at all."

---

[a] Genereaux remarked in his Oct, 24, 1982 interview with John K. Smith: "I think anybody that's been in research and then goes into application has a real advantage, because he respects the need to use the best methods. *Most of design is a compromise anyway. It takes a lot of judgement, but if you understand the principles that are in back of it, you have an edge on making better judgements*" (Emphasis added.)

The arrangement was kept as simple as possible, with just one or two pieces of equipment – vessel, agitator, centrifuge, etc, – in a cell, thus simplifying the space and layout problems.

This "kitchen" solution made feasible the parallel design and construction plan and left the Met Lab free to revise their processes whenever they saw a better way to do it, long after completion of the basic cell design, as mentioned above. This solution also avoided the necessity of dividing the design group into several different sections, one for each of the main processes under consideration, and that might eventually be selected.[52]

Separation Plant Physical Design – *Detailed Implementation*

The nuclear nature of the separation process imposed two requirements unique in the process industry of that time – avoidance of critical mass, and radiation protection. The critical mass problem was solved by proper sizing of batches and vessels.

The radiation-protection problem was more difficult; it required isolation of the process behind heavy shielding, remote sensing and remote control of the process, remote maintenance and remote replacement of equipment and piping, and precise control of the dimensions of the concrete cell structure.

Remote replacement meant standardization of equipment and precise tolerances of equipment and piping. "We designed the cell systems to hundredths of an inch, overall," Genereaux said.[53] This precision in a 13-foot wide cell would have required dimensional calculations to six significant figures, then rounding to five, a calculation like a traverse closure, using hand-cranked calculators. The cell floor was also designed to these tolerances to permit precise equipment location. The equipment was located by trunnions that fit into precise guides in the cell walls.

These precise tolerances were achieved by strict equipment specifications, rigid inspection, painstaking spool fabrication, and by use of mockup cells at Wilmington and at Hanford. The Wilmington mockup was used to verify the workability of cell design prior to drawing issue. In the Hanford mockup, each suite of equipment and piping for a given operating cell was assembled and checked. This assembly process in the mockup included the exact fitting of each piece of connecting pipe – called a "jumper" – from its connector at the cell wall to the other connector at the equipment, This jumper fitting was unique for the particular cell, meaning that jumpers were not interchangeable between cells.[54]

The connectors on each end of each jumper were DuPont-designed three-jaw mechanisms that closely gripped the pipe ends at the walls and at the equipment nozzles. These connectors were provided with a pointed hex nut powered by an impact wrench lowered remotely from the overhead cranes under the control of the human operator in the closed and lead-shielded cab on the crane. These jumpers and their connectors are seen in the cell photo on page 15.

The protection against the high-radiation environment resulted in some interesting equipment selection. Because stuffing boxes and valve seats could not be maintained without human operators, pumps and valves could not be used in the separation cells; steam and air-jet syphons were used instead. Because the plant operators were necessarily separated from the process equipment by thick concrete shielding, special process sensors had to be provided. Wobble indicators, vibration and noise indicators, and specific-gravity recorders were typical.[55]

During Hanford mockup assembly, design engineers sent out from Wilmington rectified any drawing discrepancies discovered during assembly. The Wilmington men saw to it that the rectified system conformed to the intent of the original design. Once assembled and checked in the Hanford mockup, the suite was disassembled and the components were transferred to and reassembled in the actual cell in the canyon.[56]

But here an interesting thing happened. The final re-assembly was not done by craft workers in the cell. Instead, every bit of the cell equipment and piping in the entire canyon was assembled by the overhead-crane operators using their remote-handling equipment, without ever laying eyes or hands on the equipment! Genereaux felt that if his remote-maintenance concept didn't work, it would only be a question of time until some of the cells – maybe all of them – would be permanently disabled. The remote maintenance equipment therefore and its human operators had to be absolutely proven out before hot operation, and what better way, Genereaux reasoned, than by doing all of the initial mechanical, piping, and electrical installation remotely. I asked if this potential slowdown in the construction schedule bothered Gil Church. "No," said Ray, "but the Area Superintendents questioned it." "In fact," he said, "Construction didn't even want to waste the time required for assembly of components in the mockup; they wanted to install the cell piping directly in the cells to save time. I told them 'absolutely not.' "We had to have the pre-assembly to ensure that the pipe and equipment suites were built exactly like the drawings to make remote maintenance as simple as possible. After we talked it over, Construction saw my point."[57]

Remote installation, operation, and replacement was made possible by the use of crane-mounted periscopes and televisions. It was one of the first uses of industrial television. (The as-delivered periscopes were found deficient however, and had to be sent to the Wilmington Shops for disassembly and rebuilding.[58])

As it worked out, the Area Superintendents didn't have that much to worry about. According to the Building & Facilities charts of DuPont B, it took 4.8 months to install the 221-T building equipment at 200 West, ending on September 15, 1944, twelve weeks before the T-Plant startup, allowing time for the water runs and hot testing before production started.

Certain aspects of concrete construction were also affected by the radiation problem. Construction joints were not formed against flat forms, but against corrugated sheet material to provide the labyrinth geometry required to prevent straight-line radiation streaming. The 35-ton cell covers had to be accurately fitted to prevent radiation release, so machined cast-iron forms were fabricated to ensure accurate cover geometry. Cell floors had to be absolutely plane to permit complete drainage of radioactive fluids, so a stainless-steel screed was devised with which to provide the final finish.[59]

In summarizing the reasons for total success of the hot-cell system and operation, Genereaux remarked that the three crucial keys were the Wilmington and Hanford mockups, and the remote installation of all of the cell facilities by the plant operators.[60]

Reactor Design

I have no first-hand account of reactor-design methods or activities because there were few, if any survivors of the reactor design group. According to Ray Genereaux however, the reactor-design group practices were identical with those of the separation group because methodologies were uniform throughout the Design Division.

It will be useful though, to list the principal design topics that had to be addressed immediately following the final conceptual decision on February 16, 1943 for a light-water cooled, graphite-moderated pile, with uranium fuel in aluminum tubes. (The list was abstracted from DuPont A, pp. 79-134). Finally, two special, reactor-related topics of interest are presented.

Principal Reactor Design Development Tasks

Development through final design was required for:
- The uranium coating
- The thin-walled aluminum tubes
- Reactor shielding
- Inlet and outlet water connections for the aluminum tubes
- Methods and equipment for charging and discharging uranium slugs
- Graphite block dimensions, machining methods, and assembly in the pile
- Uranium billet and slug manufacturing methods
- Pile control systems
- Pile helium system required for: excluding air from the pile; assisting in tube-leak detection; drying out the pile following tube-leak repair; cooling the thermal shield: and removing radiation-generated gases.

The special, reactor-related topics of interest are discussed next.

Who Advocated 2004 Tubes?

The literature is divided on the identity of the DuPont engineer who urged the increase to 2004 tubes from the conceptual 1500 (Fig. 2.3). Rhodes said it was John Wheeler. Graves, in his unpublished interview with John K. Smith said, "Greenie and I had agreed we had to have a factor of safety in this." Compton said. "This man was Assistant Technical Adviser in the Special Division of the Explosives Department," a description of George Graves' position on the TNX Chart, Figure 5.2. Seaborg said that he talked to several DuPont men at the time and they unanimously said it was Graves.[61] Even through the fog of a half century it seems clear that it was George Graves.

Consequences of A Failed 1500-Tube Reactor

Russell Stanton said in his interview that building a new reactor with 2004 tubes would have required 8 to 10 months after he received the drawings, assuming of course that block and tube procurement could be scheduled in parallel with foundation and building construction.

So, for 10 months we would have been limited to the ^{235}U, gun-type bombs. The problem with that was that after the one, August 6, gun-type bomb there were no more and wouldn't be until December of 1945. Given Compton's description of the strong opposition to surrender inside Japan – the circumstances included shooting a Japanese general and an invasion of the Imperial palace to steal the Emperor's surrender speech – even after two bombs had been dropped, a single bomb, with no others apparent to the Japanese for months, may have been insufficient to overcome the strong internal opposition to surrender, and the November 1, 1945 Kyushu invasion would have been carried out as scheduled, with great loss of life. [62]

Design Changes

Apparently there were surprisingly few of these. Frank Mackie maintained that design changes were "at a minimum." I asked Matthias if Wilmington, Washington, D.C., and the Met Lab sent many design changes to Hanford, and he replied, "(DuPont's) Engineering Department kept well ahead of the job. I don't know of any changes (from them). The Army left me alone entirely. The Met Lab didn't do much in the way of changes. DuPont wouldn't even build until they had approval from the Met Lab."

Russell Stanton said, "We did get a number of design changes but they were for detail items, and not for major rework. They were annoying but not critical. It was not a major problem."[63]

Average Hours Per Drawing

This would be an interesting number, but there are no data in the References sufficient to calculate it. Although the total number of drawings is known, the split of design-engineering hours among engineering/drafting, the design-development items of Table 5.2, the Wilmington engineers sent to Hanford, and the design hours expended on Clinton is unknown. Estimates of this split would be unavailing guesswork.

What is known is that total Wilmington design engineering as a percentage of total Hanford capital cost was very low, as discussed in Chapter 7.

The Wilmington Shops

These shops had served DuPont as a source of equipment manufacture and development since before World War I. As Ray Genereaux said: "DuPont had their shops a good while before 1917. Because of the competitive nature of the powder business, a good deal of DuPont's machinery was proprietary, so in order to keep these machines secret, they made them themselves in their shops. That's where the Wilmington Shops came from."

Walter Simon remarked, "We depended on the shops for our commercial plants. In the chemical business at that time you couldn't just go out and buy a lot of equipment – you had to make it yourself. They were the backbone in the '20s and '30s of our chemical equipment supply."

This quarter century of acquired development know-how prior to Hanford provided an essential advantage in development, fabrication, assembly, re-building, testing, and run-in of Hanford components and systems. The Shops could respond very quickly – much more so than going out to commercial shops – to questions from Design and therefore made a material contribution to Hanford's rapid completion. The Wilmington Shops' Hanford effort comprised over 40 items, including block-program development, testing and development of other pile components, fabrication of pile components, and fabrication, development, testing, and rebuilding of separation-cell components. The cell centrifuges and canyon periscopes, incorrectly fabricated by the vendors, were re-built at the Shops. Examples of Shop-fabricated items were the remotely-operated electrical and piping connectors, remote-handling tools for the cells, and the charge machine and shielded discharge-end cab for the pile.

In addition to this time-consuming Hanford work the Shops worked three shifts to produce equipment for other DuPont projects, as well as Government-contract war production on tank drives, gun rings, and classified new designs requiring very close tolerances.

The Shops were closed for good in September, 1993, according to Simon, because there are now enough commercial developmental and testing services available. Tepe also remarked that computer control of manufacturing has eliminated the need for developmental fine tuning of manufacturing equipment.[64]

• **Procurement** •

DuPont acquired the large quantities of unusual plutonium-process equipment over a 25 month period. Buying standard process equipment on this schedule in peacetime would have been routine. Obtaining the exotic HEW components under wartime conditions was another matter altogether. Despite these adverse conditions. DuPont placed 47,304 purchase orders worth $2.1 billion (1994) plus 74 subcontracts worth $1.1 billion (1994) in 47 states.[65]

DuPont's procurement was accomplished by two principal means:

- Amending, upon Corps suggestion, their standard procurement system with time-saving modifications
- Establishing two ad hoc groups to deal with some specific problems

Principal Time-Saving Modifications

These were:

- Reserving space in vendors' schedules by early orders based on quantity and tonnage estimates made prior to design or drawings
- Waiving competitive bidding if that would save time
- Placing verbal orders concurrently with requesting Corps approval; the Corps guaranteed compensation to DuPont for cancellation costs if the Corps subsequently disapproved a verbal order
- Splitting large orders among several vendors despite differing bid prices, if splitting would save time
- Buying from warehouse stock – rather than waiting for the less-expensive mill rolling – whenever the schedule was in jeopardy
- Procurement of used equipment to avoid the time required for procurement and fabrication of new equipment
- Buying machine tools for loan to vendors to increase their production rates.
- Permitting lower-tier orders to subsidiary vendors[66]

Ad Hoc Groups For Specific Problems

Pile components were then totally unheard of in industry and therefore required unusual procurement efforts. Accordingly, DuPont staffed two Construction-Division special procurement groups at Wilmington – the Block Group and the Uranium Slug Group. The former dealt with the laminated biological-shield blocks and eight other pile components.[a,67]

The Block Group

This group was responsible for 1245 four-foot composite cuboids – the biological-shield blocks – averaging eight tons each and consisting of alternating layers of thick steel plates and stacks of thin Masonite sheets, all secured by steel tubes inserted through holes in the plates and Masonite – then welded in place.

---

a Besides the shield blocks, the other eight "Block" items were: thermal shield blocks; inlet and outlet nozzles; graphite bars; aluminum fuel tubes; cast-iron sleeves; expansion bellows; tie straps; and the gun-barrel tubes.

The inlet and outlet nozzles comprised 283,950 parts made by 16 fabricators for assembly by a seventeenth. The nozzle production flow-monitoring chart is shown in Figure 5.8.

Tolerances of the blocks, plates, hole and tube diameters, and hole centers were held to thousandths of an inch.[68]

These blocks, comprising a total of 352,694 separate pieces, were assembled by 10 operations under 32 purchase orders by 19 vendors in 14 cities from Chicago to Long Island, and from Rutland, Vermont to Laurel, Mississippi. Components and partial blocks had to be transferred among vendors via 33 different routes, not only to save manufacturing time, but because no one vendor had complete fabricating capacity.[69]

To manage this complexity the block group, staffed with eight core members and 35 field inspectors, consulted with members of DuPont's engineering, manufacturing, inspection, procurement, and expediting sections. The group held weekly meetings, issued weekly status reports of purchase orders and current problems, and employed production-tracking devices such as the following (found in the Hagley collection).

- Flow charts of vendors, routes, operations. etc. (Figures 5.7 and 5.8)
- Completion, acceptance. and shipping charts (Figure 5.6)
- Control Cards (Figure 5.9)

The multidisciplinary nature of the group and the experience of its members enabled immediate resolution of the majority of the problems in the weekly meetings.[70] In addition to routine topics, 70 special fabrication and administrative problems were dealt with, as shown in Table 5.3.[71]

The first order under the laminated block program was placed slightly before August 19, 1943 and the last block for the 105 B reactor was shipped on March 6, 1944.[72] Allowing 30 days of this period for final fabrication, resolution of the 70 problems averaged 2.4 days each, and that time included discovery and discussion of the problem, agreement on the solution, and implementation of the solution. The 2.4 days also included the effort required for all of the routine topics.

Other Pile components

The uranium slug program and the balance of the Block Program encompassed the other exotic pile components. Eighty-two vendors and well over 129 tests and other special fabrication problems were involved in this procurement.[73] The DuPont slug-procurement report was in semi-summary form so alluded to but did not describe every problem.

Separation Plant Components

These components comprised conventional equipment, exotic equipment, and stainless-steel pipes and vessels.

The conventional equipment, such as centrifuges, agitators, and steam-jet ejectors presented no special procurement problems, as they had long been manufactured in industry. The principal problem was inspection of this equipment – then rebuilding to correct deficiencies, as in the cases of the centrifuges and the periscopes.

The exotic equipment, designed especially for Hanford was built to Design-Division order at the Wilmington Shops, as subsequently described.

Because of the relative ease of procurement of conventional equipment, and because of the capabilities of the Wilmington Shops, the separation design group did not find it necessary to set up any special procurement groups.[74]

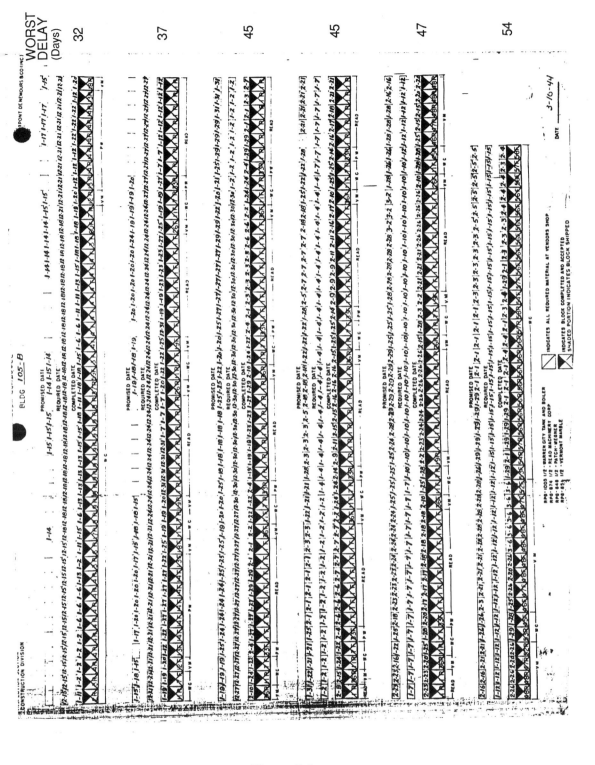

Figure 5.6

SHIELD BLOCK COMPLETION, ACCEPTANCE AND SHIPPING CHART

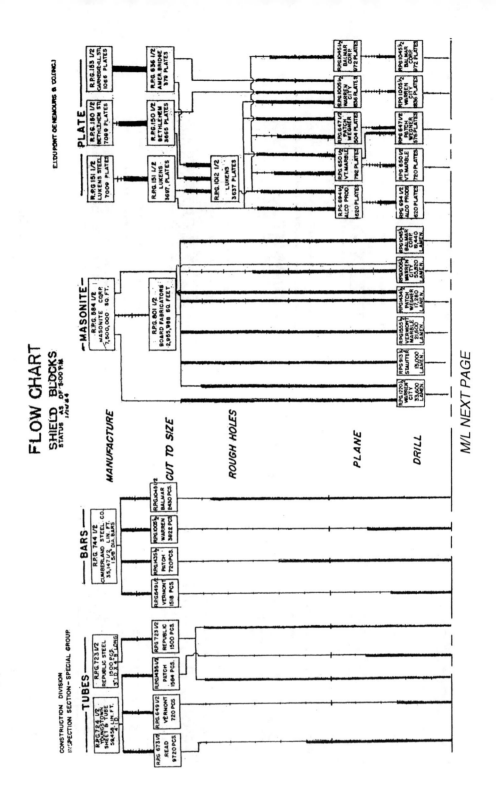

The archive copy had been pieced together from four quarter-piece photostats of the orginal. The original appeared to have been about 24 ins. x 36 ins. (61 cm x 91 cm).

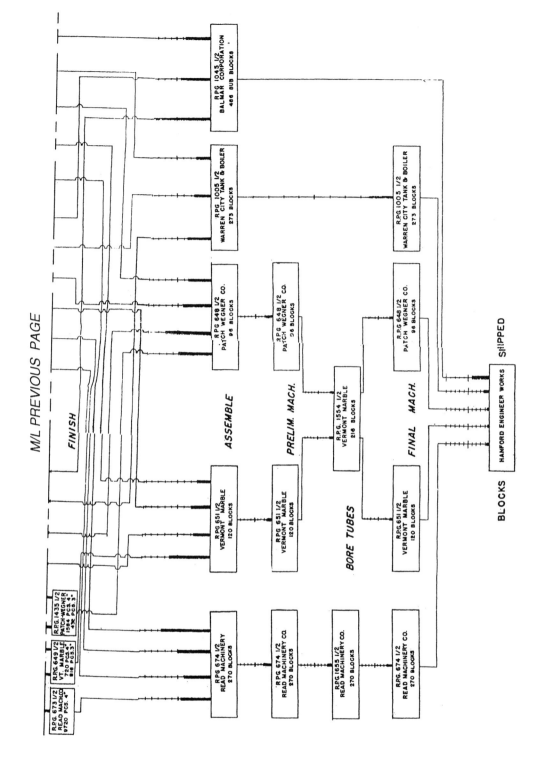

Figure 5.7

BIOLOGICAL SHIELD BLOCK WEEKLY FLOW CHART
5:00 PM Status on January 20, 1944
Tracks inter-vendor routing, vendor operations, component and block quantities,
and purchase order numbers

Figure 5.8

REACTOR NOZZLE FLOW CHART

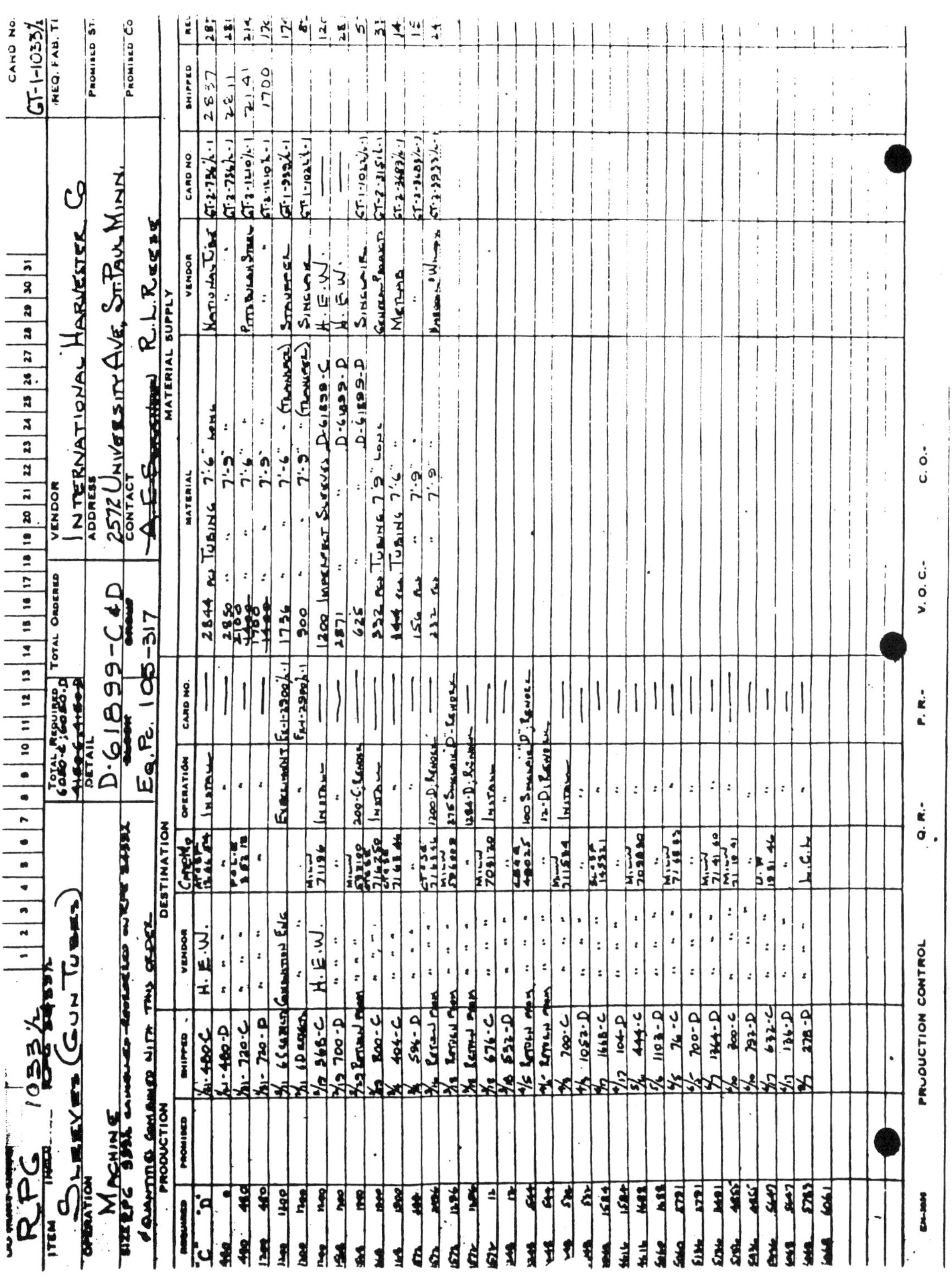

CARD NO. GT-1-1033½

RPG 1033½

ITEM: SLEEVES (GUN TUBES)

OPERATION: MACHINE

TOTAL REQUIRED: 6060-C; 6060-D

DETAIL: D-61899-C&D

GROUP: Eq. Pc. 105-317

VENDOR: INTERNATIONAL HARVESTER Co

ADDRESS: 2572 UNIVERSITY AVE, ST. PAUL MINN.

CONTACT: R.L. REESE

PRODUCTION: REQUIRED | PROMISED | SHIPPED

DESTINATION: VENDOR | OPERATION | CARD NO.

MATERIAL SUPPLY: TOTAL ORDERED | MATERIAL | VENDOR | CARD NO. | SHIPPED

PRODUCTION CONTROL | Q.R. | P.R. | V.O.C. | C.O.

Figure 5.9

CONTROL CARD FOR SLEEVES

Table 5.3

SPECIAL FABRICATION PROBLEMS – LAMINATED BLOCKS

Type of Problem	Number of Problems
Vendor availability	2
Tool availability and capabilities	9
Template, gage, and jig fabrication[a]	12
Developmental and testing operations[a]	16
Tolerances	8
Welding and weld distortion	5
Assembly	2
Develop new materials	1
Material allocations and deficiencies	4
Vendor capabilities	4
Vendor changes and cancellations	3
Drawing changes: the first block drawing was revised 106 times in four months	1
Shipping methods	3
Total special fabrication problems	**70**

[a] Most of this was done at the Wilmington Shops

The Stainless-Steel Plate Warehousing Order

In April, 1943, DuPont determined that the quickest procurement method for stainless-steel pipe and vessels for the separation plant was to buy – with design details yet unknown – a large quantity of stainless-steel plate in a variety of thicknesses, in order subsequently to supply the fabricators of pipe and vessels – after such vendors were chosen, and the pipe and vessels had been designed – with the plate. Accordingly, a purchase order was placed with a Pittsburgh firm to warehouse and inventory the plate, cut it to fabricator-required sizes, and ship it to the future pipe and vessel fabricators.

DuPont stated that this method – as compared with two other alternatives – saved three months in the schedule, but that the purchase order cost $86,600, an extra expense as compared with direct shipments from mill to fabricator.[75]

This example of schedule management illustrates a major topic in project cost, the details of which are discussed in Chapter 7.

• Hanford Field Management •

Hanford's $4 billion process plant, comprised of totally unprecedented components, was brought in a year ahead of schedule, with a nearly flawless startup, and with a cost overrun limited to 11%.

How did DuPont do this? The explanation encompassed:

- A massive and time-tested field organization, staffed with men long-experienced in process-plant construction who had worked amicably together for years
- Scheduling by the DuPont-invented Critical Path Method
- Day-to-day supervision of crafts by DuPont's Assistant Division Engineers
- An effective Quality Assurance program a generation prior to the coining of that phrase

Field Organization

DuPont's management organization at Hanford under their Field Project Manager, G.P. (Gil) Church, was very large – more than 5800 management personnel, a total commensurate with the management of a peak force of 45,000 and a construction camp that was the third-largest city in Washington.[76]

The organization consisted of a normal, but very large hierarchy with extensive branching – down to a ninth management level in the largest branches. This large number of levels was dictated by the need to supervise the 5800 while simultaneously restricting those reporting directly to any manager to a workable number, five as it averaged out.

The six DuPont organization charts for Hanford were too large to include here; they total 24 ft in length, a number that in itself emphasizes the project's magnitude.[77] As a substitute for these charts I display the organization on two summary charts.

Figure 5.10 is a simplified chart of the nine branches reporting directly to Church. In that chart I expand the Reactor Division as an example of a typical construction sub branch. Table 5.4 is a matrix of branches and branch functions versus reporting relationships, management levels, and personnel counts.

The nine positions reporting directly to Church appear unwieldy. It is possible, however, that Church had his assistant deal with the Service and Control branches. The special-assignment men probably functioned as staff and would therefore report

Figure 5.10

THE HANFORD ENGINEER WORKS
SIMPLIFIED FIELD MANAGEMENT ORGANIZATION

FIELD PROJECT MANAGER, — G.P. Church (1)
- Assistant Field Project Manger — T. L. Pierce
  - Assistant Field Project Manager, Construction — G. E. Bubb (2)
    - Assistant
      - Field Supertintendent — L. S. Grogan (3)
        - Assistant — W. E. Redmon
          - Assistant Field Superintendent, 100 Areas — L. G. Ahrens (4)
            - Division Engineer, 101 & 105 Buildings — R. C. Stanton (5)
              - Assistants — G. A. Christian & R. E. Stewart
                - Records Engineer — R. M. Kenady (6)
                - C. A. Lyneis (7)
                - 101 Coordinator — J. L. McIntire (6)
                - Inspection — H. S. Cline (6)
                - W. R. Hardleben (7)
                - 105-B Coordinator — R. F. Stewart (6)
                - J. F. Mitchell (7)
                - C. P. Sparks (7)
                - 105-D Coordinator — E. B. Rayburn (6)
                - G. H. Sowers (7)
            - Division Engineers for 100-B, 100-D, 100-F (5)
          - Assistant Field Supertintendents (4) for: Crafts;[a] 200, 600, 800, & 900 Areas; Instrumentation, & Camp Layout
      - Field Superintendent — R. K. Mason for 300, 700, 1100 Areas (3)
    - Service Superintendent — E. L. Pleninger (2)
  - Assistant Field Project Manager, Safety — F. H. Mcdonald (2)
  - Assistant Field Project Manager, Control — G. E. Hillman (2)
    - Hanford Housing Superintent — B. M. Taylor (2)
- Security Agent — H. F. Highsmith (2)
  - Assistant Field Project Manager, Engr. Admin. — W. V. Krewatch (2)
  - Assistant Field Project Manager, Special Assignment — C. H. Trask (2)
    - 100 and 200 Areas
  - Special Assignment — E. M. Elliott (2) & C.H. Trask (2)

Stanton's Division expanded to show typical detail to 7th management level. Other divisions are similar.

---

[a] The large, intraplant bus system (Ch. 2) was operated under the Transportation "craft" in this branch.[78]

Table 5.4

HANFORD CONSTRUCTION ORGANIZATION
BRANCH FUNCTIONS & MANAGEMENT RELATIONSHIPS

BRANCH & FUNCTIONS	NO. REPORT'G UP TO BRANCH HEAD	AVERAGE NO. REPORTING UP IN BRANCH	LOWEST LEVEL OF MGMT.	COUNT OF MGMT. PERSONNEL
Field Project Manager	9	—	1	10
Construction Construct all areas except Richland Village	3	5.4	7	840[1]
Service Employment, recruitment, statistics, reception, terminations, training, investigations, medical patrol, fire, civil defense	4	3.2	9[2]	2550
Safety Safety management in all areas, including Richland Village	2	NA	4	20
Control Railroads, office management, clerical, mail communications, purchasing, receiving & stores, payrolls, paymaster, payroll accounting timekeeping	2	3.2	9	1420
Housing[3] Housing, activities, feeding, barracks reception, rationing, utilities engineer, trailer camp	6	NA	4	70
Security (Organization not provided in charts)	—	—	—	—
Engineering Administration Job improvement, cost analysis, cost engineering, estimating, reports & charts, reproduction, records, specifications, inspection, contracts, history, field design, industrial engineering, staff engineering	5	4.4	9	390

Notes:
1. Augmented beyond chart, based on Stanton interview re: 105 Bldg.
2. Estimated on basis of "Control"
3. Original chart may be incomplete

only intermittently, thus limiting Church's day-to-day contacts to five subordinates, a manageable number.

The Housing branch shown on the original chart counts to but 70 management people, a questionably small amount for dealing with Hanford's huge camp and the Richland village. I had, however, no rational basis for augmenting the number.

An Accounting Department appears nowhere in the original charts. A separate DuPont tabulation, however, found in a project accounting file at the Hagley shows 551 accounting personnel in July of 1944, the date of the overall charts. I have included this number in the total 5800 stated above.

The Scheduling function is also omitted from the original charts. I determined in a phone conversation with Russell Stanton, the Division Engineer for Reactors, however, that this function was accomplished at two levels: the Division Engineers' and in the office of the Field Superintendent, L.S. Grogan. Scheduling personnel would have been a relatively small number compared to the project total, so I have not added any quantity to the total.[79]

It is clear from an examination of these charts that DuPont provided for all of the routine administrative tasks to be found on every large construction project; it could not have been otherwise. The very facts of the magnitude of the project, the enormous work force, the rapid completion, and the trouble-free startup testify to an experienced, effective, and tightly controlled organization.

As noted previously, Matthias remarked that "DuPont cut no corners." On office routine, Simon confirmed the large size of the field management forces mentioned above: "We had a large amount of man and woman power to take care of these massive routine chores."

DuPont's Invention Of The Critical Path Method (CPM) And Their Use of it At Hanford[80]

DuPont invented and used CPM about 1940, two decades before they announced it to the contracting world in the late 1950s.

There was no single inventor of CPM at DuPont, according to Russell Stanton, and no one project on which it was first used. The method evolved over approximately three years prior to Stanton's arrival at Hanford.

As the proto concept began to spread throughout the company, many meetings were held to answer questions on the method, and a great deal of correspondence occurred as the concept came into use on various DuPont projects. Everyone saw the potential advantages of the method, and Engineering, Construction, Procurement, and Expediting made contributions from their individual points of view.

The development chronology was approximately as follows, as nearly as can be determined after six decades:

1937 to Sept., 1939. Stanton was assigned as a construction engineer at DuPont's carbon tetrachloride facility at Richmond, Va., and at their ethyl chloride facility at Baton Rouge, La. He remembers that CPM did not exist on either of these projects.

Sept., 1939 to Jan., 1941. Stanton was assigned to the adipic acid facility at Belle, W. Va. Part way through this assignment, the DuPont staff at Belle began to develop and use the initial phases of CPM. They began by connecting the bars on bar charts – the tails of prior bars to the heads of the next bar in the construction sequence.

We have to assume that DuPont engineers at other facilities were then beginning to work on the concept simultaneously with those at Belle, because Stanton remembers that CPM was discussed throughout the company. And if we wish to assign a date to the origin of CPM, it might be 1940, which is "part way through" Belle.

Jan., 1941 to May, 1942. Stanton was at the ammonia unit at Morgantown, W.Va.

Here, he remembers, they were still in the early phases of CPM, but by this time it had taken on a number of the features of the final CPM.

It is important in the validation of this history of the development of CPM to note that Matthias recalls seeing DuPont's CPM in 1942, and told me about it in April of 1993, five months before I discussed CPM with Genereaux and 22 months before Stanton recited this chronology. I quote from Item 29.FM of his interview:

"I had learned about how they'd do that because Groves at one time had me review some of the construction jobs they were doing, long before he was even with the Manhattan District." (That is, before Sept. 17, 1942 when Matthias and Groves were in charge of construction of the Pentagon.) "And he got me to review some of their ways of making reports, because they were so far ahead of any of the other builders in the country – he just wanted to know how they did it. So I had gone through that kind of drill on my own – that's one of the personal assignments Groves gave me."

"They – for the first time – they didn't talk about it then, but they were using the program of critical path."

May, 1942 to Oct., 1943. Stanton had three other assignments, one of which was at the SMX unit at Clinton, where he assisted in the building of that unit and running cooling-system hydraulics experiments on it.

By the time he arrived at Hanford in October of 1943, DuPont had developed CPM to the final stage that the rest of us learned about in 1959: nodes; branching activity lines; parallel activities; the longest-duration sequence being the critical path; float in the parallel, non-critical activities. DuPont at that time was making formal CPM drawings, hand-drafted of course, on all of their projects, including Hanford.

Matthias, Stanton, and Genereaux differ as to whether the term "Critical Path Method" was used by DuPont at Hanford – but they do agree that, regardless of nomenclature, it was the same CPM to which the rest of were introduced in the late 1950s.

Stanton said that at no time did DuPont instruct their men to keep CPM secret. He is still puzzled that the concept took so long to become general knowledge, because DuPont's engineers often talked to outsiders about the concept.

They would regularly point out to particular vendors or subcontractors that they were becoming bottlenecks in DuPont's schedules, and when these people asked how they knew that, then DuPont would show them DuPont's CPM charts to demonstrate the critical path, and how the vendor was holding them up.

CPM Use At Hanford

Hanford CPM was managed by the Construction office, with the Wilmington Design Division going in parallel with Construction's CPM. "We would start out with the first issue of the schedule," Stanton said, "which was based on Wilmington's design schedule. We would then notice problems with Wilmington's schedule and ask if they could do so-and-so. If they could, they would accomodate us but sometimes they couldn't because of lack of design information, or something, so we'd propose something else, and we'd work out the final first issue that way. But our Field office always had the initiative in making the schedule." Referring to Figure 5.10, the Central CPM was kept in the office of Field Superintendent Grogan, with the Division Engineers, such as Stanton, keeping their own.

"We at 105 were moving so fast we didn't have time to notify them in Central of our changes," said Stanton. "We'd carry the changes along in our heads until we came to one that would affect the final completion date. Then we'd get with Central and work out what we and the project as a whole would do, and then the CPM would get updated." (As a commentary on the fast pace of pile construction, Stanton

The 101 Building

A construction facility immediately north of the Hanford Camp. Here the graphite blocks were stored and machined – then laid up in trial arrangements in order to verify the actual location of each block in the reactors.

HANFORD SCIENCE MUSEUM

---

remarked, "The Met Lab couldn't believe we were putting the parts together as fast as we did.")

Matthias said, "I've always thought that that CPM system of DuPont's was what led us to do a very efficient job at Hanford."

Engineer Supervision Of Crafts

As shown in Figure 5.3, the Construction Division's General Manager in Wilmington reported to the Chief Engineer. This relationship was reflected throughout DuPont down through each project to the foreman level; the foremen on all DuPont projects reported for technical direction to the cognizant Assistant Division Engineer, and had done so for many years. (Supervision of personnel matters was by the craft superintendents.)

Matthias was the first to tell me about this system: "The engineer would lay out every task for a day ahead. Then the engineer would supervise it. The workers and the foremen were told just what to do by the engineer."

Russell Stanton provided the details of the system. "The engineer-managed system was prevalent throughout DuPont and they had been doing it as long as I had been there. The engineer in charge of each area would have to keep his own notes of what was to happen next, which would include sketches and schedules. It was pretty informal – the engineers and the crafts would see each other all day long, so he would hand a copy of his notes, sketches, and schedule to whomever was in charge – the foreman or the superintendents. The usual medium was either the engineer's handwritten notes or oral instructions and they would describe the work in that area for the next one to three days."

"Occasionally there would be a reason for typed instructions but mostly it was the engineer's handwritten sketches and notes, or oral instructions. These instructions were supplemented by gang-box safety meetings with the craft foremen. At the start of each new construction phase the engineers would hold a Job-Instruction meeting with the foremen to describe the objectives of that phase."

"For occasional, important changes the engineers and secretaries would work at night to issue written material to foremen. After the written or oral instructions, the engineers were constantly on the job to see that things were going right, and would instruct the craft foremen when necessary."

"The engineers also directed reactor-related activities away from the job site. We had several 105-Area shops – the 101 building at the Camp area, and the tube shop, the Masonite shop, and the plate shop at the White Bluff shops. My people would expedite things through these shops."

According to employment data provided by Stanton, there were about 90 foremen at the reactor building and 15 to 20 engineers out in the work areas, or a little over five foremen per supervising engineer.

I asked if the craft unions objected to the engineers directing the foremen. "No," Mr. Stanton replied, "they asked for directions from the engineers and were appreciative of this help."

Concerning the engineers' planning for shortages of manpower and other resources, Stanton said, "I can not think of a single case of schedule slippage due to lack of manpower at the 105 building. We did have to make sure that we planned around the available drawings, materials, and manpower. In other words, we made sure by looking ahead and planning for it that we never got caught short.

Stanton is essentially correct about his maintaining the construction schedule. According to the progress charts of DuPont B, the overruns on the forty-week schedules for the three reactors were 4.7%, 13.8%, and 16.3% for Reactors B, D, and F respectively, which were relatively minor slippages.

Matthias appreciated the effect of this management method. "I think," he said, "that as far as the overall efficiency, it was all because the crafts were guided in what they did and told what they had to do, and what part of it, and everything laid out for them every minute by the engineering group. I think that was the difference."[81]

Quality Assurance (QA)

It appears that Dupont had a thorough understanding of the need and the procedures required for QA, even though they didn't then call it that. Their meticulous care in engineering and checking has previously been described, as have the pains they took with separation-plant mockups at Wilmington and Hanford and, as mentioned below, the pile-layup mockup in Hanford's 101 Building. Both Tepe and Stanton comment extensively on DuPont QA in the subsequent paragraphs.

DuPont's Overall QA

During the interview with Jack Tepe, I asked how DuPont achieved the quality they did without having a formal QA program. He remarked as follows:

"We had QA. We didn't call it QA, but we had it. For example, every material we bought was tested to make sure it met the specification. We used enormous quantities of stainless steel and exotic alloys and they were all tested."

"Construction had always had a major inspection capability, They always had inspectors in all the vendors' shops, and when critical machinery was shipped they had inspectors ride the caboose of freight shipments to make sure they weren't bumped, or lost, so we had QA but we didn't call it that. We had administrative paperwork for all this but it was a minimum, and we didn't talk about it."

"There are a lot of things that we're putting labels on that are not new."

Hanford reactor QA — Preparation

Prior to building Hanford's piles DuPont had seen the need for a total physical study of the graphite blocks, the methods required for their layup, and many other pile developmental studies. They therefore built the SMX unit at Clinton and experimented with it for many months before actual layup started at Hanford. Stanton had been assigned to the SMX experimental work, no doubt because DuPont had selected him to manage reactor construction at Hanford, and had participated in the construction of the SMX, followed by participation in the SMX experimental program.

His SMX experience indicated, among other things, the need for precise and detailed written pile-construction methods. "We thought through the whole construction problem of the pile at the start," he said, "and decided that we had to do something to ensure that it got done right. We got out there before there were drawings to work from,[a] so we took the opportunity to plan the entire operation. We'd work well into the night writing procedures for assembling the pile. It was a form of recreation because there was really nothing at all to do at night in the first months."

"We wrote between 200 and 300 procedures, averaging over ten pages each, so I would say we had well over 3000 pages of procedures."[b] (At least 10 three-in. binders.)

"One procedural problem was to keep the graphite from being contaminated. It was absolutely crucial from the nuclear-physics point of view that the graphite be kept absolutely clean. So we had a complete set of procedures written covering all this. For example, we knew we had to keep all our work clothing absolutely clean to avoid contaminating the graphite from the clothes – the gloves, coveralls, shoe covers, etc. So we had a laundry procedure that specified what soaps and detergents could and could not be used." Examples of some other procedures – furnished by Stanton – are shown in Table 5.5.

Those construction-QA procedures involving design considerations (as opposed to purely construction methods) were requested, then reviewed by Wilmington and the Met Lab. An interesting example was the basic pile-erection procedure. The Wilmington Explosive-TNX Design Group advocated a layered approach, based on layers of the eight-ft shield blocks. In this proposed method the bottom course of biological shield blocks (together with inner layer of thermal blocks) would be laid in place and welded. Then the gun-barrel sleeves and doughnuts would be installed concurrently with layup of the first eight ft of graphite-block layers: thence in successive eight-ft layers until completion.

Stanton and his engineers, however, saw nothing but trouble in this sequence. The inevitable conflicts between graphite-block workmen and engineers, and the ironworker-weldor-engineer team installing the shield blocks would slow down the

---

[a] 105-Building process drawings were 2% complete on Oct. 1, 1943, according to the Design Status chart of the 105 Building, in DuPont B.

[b] During the course of construction, unforseen difficulties and problems would arise and Stanton's group would then write additional procedures to prevent recurrence of these difficulties.

Table 5.5

CONSTRUCTION PROCEDURES
– Some Examples –

1. Pile Erection

2. Graphite-block manufacturing procedures to assure dimensions to thousandths of inches (hundredths of millimeters)

3. Block-shop equipment maintenance procedure. Equipment had to be kept in precision condition without lubrication or other foreign-material contamination.

4. Block-shop equipment set-up and operation

5. Block-shop equipment inspection (go-no-go limits)

6. Block handling and storage
   - Segregation of graphite into the three quality grades by heat number and vendor
   - Maintaining graphite-grade identity from box car to location in pile. (The Met Lab had specific requirements for locations in the pile for each of three grades.)

7. Clothing types for graphite-block handlers, block manufacturers, and block erectors

8. Aluminum-tube testing: pressure; straightness; land position, size, and orientation

9. Steel plates for shield blocks: fabrication; fit-up; cleaning; identification

10. Masonite for shield blocks: fabrication; fit-up; cleaning

11. Helium leak testing at reactor: welds on perimeter foundation pan; side-shield welds; other reactor welds

12. Graphite layup procedures

13. Welder qualification, testing, and identification

14. Equipment functional testing procedures

15. Security procedures: classified document control; isolation of areas; protection of parts.

There were many more than these examples; they totalled well over 3000 pages.

job and risk defective quality of both shield blocks and graphite blocks, in the opinion of Stanton's group. Neither the Construction nor Design Divisions would budge and the decision had to be bucked up to E.G. Ackart, the Chief Engineer, who decided in favor of the Construction Division.

A photo exists of the graphite-block workers beginning at the bottom of the deep cavity formed by the shield blocks assembled to their full height.

Hanford Reactor QA — Construction

Construction QA "keyed on my chief inspector, H.S. Cline," said Stanton. "He was responsible for the development of all the special gages, other measuring devices, and procedures that were required for manufacture, checking, and assembly. He saw to it that all these operations were done exactly as specified at the 101 Building, the White Bluffs Shops, and at the reactors themselves.[a] He kept complete and permanent records of all the manufactured parts and their testing, and the conformance to tests, and he kept them by part name and number."

"We had a records chief, Reed Kenady, who tracked the receipt and handling of graphite blocks and all the other parts. He'd meet the box cars at the warehouse, check off all the items as they were unloaded, record receipt of the item, monitor the warehousing, and record the warehouse location for each piece by name and part number."

"After the trial layups in the 101 Building,[b] he would secure design instructions for graphite-block locations in the pile and prepare work orders for their installation. Previously he had done work orders for processing the groups of various qualities of blocks, and his final work orders for block installation ensured that the blocks of the right configurations and qualities got installed in exactly the right places."

"Complete paper records were kept of all of the above, and the Met Lab would come out periodically to review these records to verify that the pile was being assembled correctly according to scientific requirements."

All welds, according to Stanton, were inspected for 100% of their length. The reactor envelope welds were inspected by a soap test, using a vacuum box with a glass window. Pipe welds were X-rayed. Shop welds were inspected either by X-ray or by dye penetrant. Welds were stamped with the weldor's number, except for the reactor envelope welds, for which written records were kept showing locations of each weldor's work on the envelope.

"Finally," said Stanton, "an overall test was made of the completed reactor envelope by pressurizing the entire block with air. All non-test personnel were cleared out of the building to ensure silence, and the test personnel then walked around the cube listening for leaks. Leak repairs consisted of re-welding the shield-plate welds and tightening the gun-barrel closure seals until no leaks could be heard. This test and the resulting repairs required about one week, and the first such test occurred one or two months prior to B Reactor startup."

"Records were kept of every weld inspection as to location, leak found, and correction done, and were signed by the inspector. The Met Lab would periodically review the weld records to verify format, contents and quality of records."

---

[a] e.g.: An Invar tape with a micrometer on one end for checking pile-block layout; jigs and fixtures for production-line measurement of length of each graphite block to design specifications – 0.003 in. (0.076 mm); squareness to 0.002 in. (0.051 mm); procedures and jigs for boring the tube hole in each block to 0.002 in. (0.051 mm); procedures for maintaining all block-shop equipment in precision condition without lubrication or other foreign-material contamination.

[b] Each block was then marked with a metal stamp to identify its location in the pile.

Hanford Separation-Plant QA

Separation-plant QA was partially described above in the paragraphs on DuPont's Engineering Practices, Cell Design, and Separation-Plant Design. The 100% checking, the mockups, and the detailed monitoring of vendors were all elements of QA.

The Hanford mockup shop operated with a full-fledged QA procedure including records of materials, fit-up, and sequencing for installation. All of these records were recallable if problems arose.

The separation mockup QA was much more complicated than that at the 105 Buildings because the differences between cell systems required a greater diversity of records. A substantial mockup QA organization was required because progress in the mockup shop was on the critical path of 221 Building progress. The mockup-shop procedures transitioned into similar procedures for actual in-cell erection.

Perspective On QA

DuPont was not unique, of course, among American firms in quality workmanship. Most manufacturers then gave thought to quality or they were out of business. Boeing, Caterpillar, and LeTourneau were three firms of that time that come to mind as quality manufacturers and, among builders, the Six Companies and Consolidated Builders Inc. that built Parker, Hoover, and Grand Coulee dams. DuPont's QA program at Hanford then, was just one example, noteworthy to be sure, of industrial quality management.

QA succeeded at Hanford for the following reasons:

- DuPont had a background in quality work in their previous munitions and chemical plants. Nylon and neoprene didn't just happen.
- DuPont was motivated towards the highest quality at Hanford because the fear of catastrophic failure was uppermost on the minds of DuPont's Board of Directors.
- Hanford QA was in the hands of the best engineers and construction managers in DuPont.

Finally, NQA-1 probably resembles DuPont's Hanford program because ANSI and ASME thirty-five years later must have canvassed all who had been major players in nuclear engineering and construction, including DuPont, and incorporated their experience into NQA-1.[82]

Quality Of Hanford Management

The plutonium veterans – Matthias, Simon, Genereaux, Tepe, Stanton, and Seaborg – pointed out that DuPont assigned their best men to manage the Hanford project. These men had been with DuPont for many years, knew each other's capabilities, and also knew the capabilities of men at other DuPont facilities and could select from these others at will. As Simon and Stanton said, "We had the cream of the crop." If they asked for a particular man, they got him.

Simon also remarked, "I know of no other operation in DuPont's history that was as heavily manned as that one, or about which the Board back in Wilmington was more fearful that something would happen that would absolutely destroy the company."

"Walter Carpenter said time and again, 'You've got the whole company in your hands.'" Then Simon added, "when I was in my thirties those things didn't bother me; now I'd be scared stiff. I can't believe what I went through one day at a time!"

To develop good construction managers DuPont had, in common with all the contractors of that time, trained their managers on previous projects under the guidance of those of greater experience. DuPont had neither a manual of construction management nor a construction management training course. Books on construction management did not exist.[83]

The Measure of Hanford Management

The first gage is that HEW worked – a nearly flawless start. Secondly, it was on time, thus avoiding the necessity of an invasion of Japan. A third view was provided by an engineer in DOE's current Hanford Site Infrastructure Office. He said in 1993 that whenever he has to go into the now-abandoned 100 Areas, how impressed he is with the quality of the workmanship there.

Seaborg's was a fourth, independent view: "DuPont did a marvelous job at Hanford," he said. This was a particularly impressive comment because Seaborg had had to sign off on all of the separation-plant operating procedures, in effect guaranteeing the separation processes. Had DuPont done a poor job, the resulting deficient separation systems would have seriously reflected on Seaborg.[84]

Chapter 6

THE METALLURGICAL LABORATORY EFFORT

The Metallurgical Laboratory – a code name to conceal its true purpose was established on December 6, 1941 at the University of Chicago to provide the scientific and conceptual bases for the plutonium program. It comprised scientists and engineers from universities and corporations across the nation, was under contract to the S 1 Committee, and was given four primary goals:[1]

- Design the plutonium bomb[a]
- Design a method for producing plutonium by irradiating uranium
- Design a method for extracting plutonium from the irradiated uranium
- Accomplish the above in time to be militarily significant

These goals were implemented by efforts under the following tasks:[2]

1. Develop the applicable nuclear physics
2. Develop reactor theory and conceptual scope, as input to DuPont's physical design
3. Provide testing, analyses, and development for graphite, graphite machining, aluminum tubes, and uranium slugs. The Met Lab was the central coordinator for the slug program
4. Complete the development of plutonium chemistry begun by Seaborg, Kennedy, Wahl, and Segre at the University of California in 1940 and 1941
5. Determine the fission products to be produced by the piles, and their chemistry
6. Develop the separation process theory, conceptual scope, and flow sheets as input to DuPont's physical design
7. Determine the effects of radiation on materials and reagents
8. Determine plutonium's metallurgical characteristics as support for bomb design
9. Study fast neutron reactions as support for bomb design
10. Develop health physics
11. Manage and operate the Clinton pilot reactor, pilot separation plant, and the SMX unit
12. Review and approve DuPont's physical drawings for pile and separation plant
13. Monitor Hanford construction

---

[a] In February 1943, 14 months after establishment of the Met Lab, the responsibility for design of the bomb was detached and assigned to the group at Los Alamos.[3]

The scientists and engineers addressing these tasks were organized as shown in the Simplified Organization Chart, Figure 6.1.

The above task list shows two principal branches of the development program – plutonium chemistry and fission physics. The latter got off to an earlier start – December of 1938, as compared to December of 1940 – and was thus in a more advanced state of development in 1942. Fermi and Wigner were therefore able to resolve the final pile concept with DuPont on February 16, 1943 – a cubical pile, light water cooled, graphite moderated, with canned uranium slugs in aluminum tubes.[a]

In early 1943 however, finalization of the separation concept was over a year in the future. When Seaborg left UC for the Met Lab in April of 1942 he had made a beginning on plutonium chemistry and had established the oxidation reduction method as one possibility for isolating plutonium, with rare earth fluoride as the carrier,[5] but a great deal of chemistry was yet to be developed.

As James Franck, Director of the Met Lab's Chemistry Division, put it, "All the qualities of our product had to be studied, first with tracer amounts and later with micrograms of the material; the fission products and their chemistry had to be developed; and radiation chemistry had to be built up from scratch. All that had to be done by men who had, with the exception of a few leaders, no experience in the field of radioactivity . . . That was no minor task".[6]

Twelve alternative separation methods were considered by the Met Lab during the first two years of its program, as follows:[7]

1. Electrolysis
2. Volatilization (dry fluoride)

   Precipitation:
3.   Peroxide
4.   Sodium uranyl acetate
5.   Bismuth phosphate
6.   Iodate
7.   Wet (lanthanum) fluoride
8. Adsorption
9. Crystallization
10. Extraction between two phases
11. Extraction of uranium, leaving plutonium behind
12. Solvent partition

Lombard Squires has pointed out however, that some of these alternatives were not bases for extensive studies. The main emphasis was on lanthanum fluoride and bismuth phosphate.[8]

Although much was learned by the use of plutonium salts in the three years of ultra microchemistry at UC and at the Met Lab, the processes so determined had to be confirmed by working with milligram quantities of plutonium metal, and that was not available until Clinton was able to begin shipments from its pilot pile and separation plant on January 1, 1944.[b]

---

[a] Other pile alternatives considered are shown in table 5.1

[b] At that time, the 200 Area drawings were 36% complete and the start of hot-cell equipment installation was only four and one-half months away.[9]

Figure 6.1

THE METALLURGICAL LABORATORY
SIMPLIFIED ORGANIZATION CHART

DIRECTOR - A.H. Compton
- Associate Director - S.K. Allison
  - Assistant Director - N. Hilberry
    - Chief Administrative Officer - R.L. Doan
    - Nuclear Physics Division - E. Fermi, Director
      - Theoretical Physics - E.P. Wigner, Section Chief
    - Chemistry Division - J. Franck, Director
      - Chemistry of Final Products - G.T. Seaborg, Section Chief
      - Ames Project - F.H. Spedding, Director
      - Section Chiefs for radioactive chemistry, byproducts chemistry, and analytical chemistry
      - **Contracts** with Iowa State, University of California, Notre Dame, Princeton, Washington University
    - Health Division - R.S. Stone M.D., Director
    - Technical Division - C.M. Cooper,* Director
      - Sections for: Development Engineering, Chemical Engineering, Separation Processes, Metallurgy & Metal Fabrication, Control & Instrumentation
      - **Contracts**, Metallurgy & Metal Fabrication: Iowa State; Battelle Institute; U.S. Bureau of Mines; Grasselli Labs of the DuPont Co.; Westinghouse; Univ. of Wisconsin.Contract: New York University
      - **Consultants:** J. Chipman; W.K. Lewis; W.H. McAdams; H.N. McCoy; R.S.Mulliken, J.R. Oppenheimer; L. Szilard; E. Teller

According to Compton (p.83) the Met Lab comprised 2000 people at its peak.

* On loan from DuPont, Wilmington

Reference: End Note 4

Glenn T. Seaborg

At the University of California, Berkeley in 1941, the year he, Kennedy, Wahl, and Segre determined Pu 239 to be fissionable with slow neutrons.

LAWRENCE BERKELEY LABORATORY/BATTELLE PRESS/ THE PLUTONIUM STORY

Arthur Compton

Project Director of the Metallurgical Laboratory, with Col. Matthias

HANFORD SCIENCE MUSEUM

Enrico Fermi
Director, Nuclear-Physics Division of the Metallurgical Laboratory

LAWRENCE BERKELEY LABORATORY/BATTELLE PRESS/ THE PLUTONIUM STORY

It is not surprising therefore, that a Clinton meeting on February 4, 1944[a] listed 30 unsolved separation-process problems.[10]

By March 6, 1944 however, Seaborg listed only 12 principal problems, seven of which were studies of variations of the main process.[11] By March then, the main process had been confirmed, with only four principal exceptions: the April, 1944 changes in the 231 purification building due to relaxation of criticality requirements; the June 20, 1944 extraction process design changes; the July, 1944 mechanical isolation changes in the 231 building; and the September 12 change to air jets from steam at the 231 building.[12]

The final separation process steps are shown in Table 2.1

The Met Lab DuPont Relationship

Nuclear physics and radiochemistry in 1942 were the exclusive province of a relative few, mostly academics at several universities, worldwide. For that reason, DuPont, having no nuclear experience, understood the need for, and appreciated the Met Lab's contribution. Walter Carpenter pointed out after Hanford's completion that it was the Met Lab's research that had enabled DuPont successfully to complete Hanford.[13] Walter Simon, in commenting on DuPont's strong capabilities in research, engineering, and design said, "Now all we did on nuclear was add the theoretical people in Chicago to that stream."[14]

---

[a] At that time, the 200 Area drawings were 51% complete and the 221 Cell Building superstructure had been underway for two and one-half weeks.[15]

Many at the Met Lab were cognizant of DuPont's process expertise. Seaborg, in a November, 1942 diary entry noted: "At our Research Associates meeting we discussed the decision by DuPont to play a central role in the production aspects of our project. We are all quite pleased because we have learned to respect the abilities and to like the DuPont men already assigned to the Metallurgical Laboratory. We have established a very good working relationship with these chemical engineers." [16]

Some Met Lab scientists however, resented what they saw as DuPont's intrusion into "their" project. These men considered DuPont to be unnecessary. One scientist remarked that if you were to give him a file and a helper, he could build Hanford himself.

These men also regarded DuPont's formal and rigorous design/construct system as much too time consuming, and their engineering conservatism as leading to a waste of money. (These attitudes were obviously the reactions of men unfamiliar with the realities of engineering and construction.)

Some of these scientists also resented having to review DuPont's process drawings. They considered themselves reduced to the role of drawing checkers.[a]

When I asked Simon if these reports of some scientists' attitudes were exaggerated, he said, "No it wasn't exaggerated, it was a conflict of honest opinion. DuPont, they thought, was making a mountain out of a molehill, and overdoing it."[17]

The day to day working relationships however, appear to have been very effective. Seaborg's many diary notes about engineering coordination between the Lab and DuPont's Squires, Greenwalt, and Cooper, and the Met Lab's monitoring of pile construction mentioned in Stanton's interview give a distinct impression of smooth cooperation.

---

[a] This is reminiscent of the attitude of a theoretical mathematician who was dragooned into working at Los Alamos during the war. When he found himself doing applied mathematics, he wrote to a friend, "You'll never believe this, but I am working problems with actual numbers. Some of them even have decimal points."

Chapter 7

THE COSTS OF THE HANFORD WORKS

A number of topics associated with Hanford costs are discussed in this chapter in the following order:
- The 1945 recorded costs
- Cost escalation to 1994, and selection of a suitable cost index
- What are the escalated costs?
- How did the HEW and total Manhattan costs compare with the magnitudes of the 1944 U.S. and German economies?
- Did coping with Hanford's unprecedented difficulties and crash schedule result in unusually high management and engineering costs?
- Project cost vs. project duration

The 1945 Recorded Costs

These recorded costs are shown in Table 7.1 and are the basis for the conclusions of this chapter; their authenticity must therefore be established. As pointed out in Chapter 4, all of these costs and the detailed records on which they were based were currently and constantly monitored by on-site U.S. Government auditors, enabling the Comptroller-General of the U.S. to testify before a post-war congressional hearing that the costs of the entire Manhattan Project were proper and accurate. The Table 7.1 recorded costs may therefore be considered a firm and accurate basis for the cost discussions and conclusions of this chapter.

Cost Escalation to 1994, and
Selection of A Suitable Index

One method of determining the 1994 costs of the original Hanford project would be to prepare a complete estimate from the original drawings, using 1994 unit prices; this is obviously not a practicable means because of the cost of the effort. Furthermore, 1994 prices of the obsolete equipment used at HEW a half century ago do not exist. The alternative of re-designing the plant to match available equipment and to suit current conditions would not only be costly, the resulting design would vary greatly from the 1945 HEW. One such current condition would be the difference in the very large and expensive construction camp, which was necessitated by the extreme, wartime gas shortage. In 1994 a large proportion of the potential camp residents would have commuted from the surrounding towns, thus making the 1945 and 1994 estimates non-comparable. It is thus apparent that even if we could afford a detailed re-estimate, its comparability with the original would be questionable.

Table 7.1

HANFORD ENGINEER WORKS
ESTIMATED AND INDICATED COSTS

	TENTATIVE ESTIMATE AUGUST 3, 1943 [a]	INDICATED COST AS OF DECEMBER 31, 1945
MANUFACTURING		
100 – Areas	$99,083,114	$92,791,269
200 – Areas	45,871,519	53,750,693
300 – Areas	2,607,848	7,122,992
500 – Outside Electric Lines	2,620,509	3,390,878
600 – Roads, Railroads, etc.	13,069,272	19,391,692
700 – Administrative, Maintenance Area	2,849,676	3,086,323
800 – Outside Overhead Pipe Lines	1,107,577	1,656,573
900 – Outside Underground Pipe Lines	15,587,581	10,709,128
Spare Parts	1,030,618	— —
Site Work and General Grading	3,010,963	2,004,898
	$186,838,677	$193,904,44
RICHLAND VILLAGE	32,946,490	34,834,28
CONSTRUCTION FACILITIES		
Administrative Area – Hanford	— —	2,491,865
Construction Camp – Hanford	17,660,000	14,237,045
Steam, Water, Sewers – Hanford	— —	11,651,477
Camp Operation – Hanford	(In Field Expense Below)	18,368,973
Temporary Construction Equipment	6,873,061	6,767,741
Major Construction Equipment	8,600,000	3,518,342
Small Tools	2,450,000	291,725
Handling Excess Materials	— —	2,343,291
Excess Materials	— —	237,513
	$35,583,061	$59,907,97
COMMERICAL FACILITIES (Bldg. & Eqpt.)		
Hanford Area	1,256,741	1,470,797
General (Including Alterations to Existing Houses)	450,000	627,409
Camp 300 Area – Military Police	— —	490,400
	1,706,741	2,588,60
FIELD EXPENSE		
Fire and Police Protection, and Sanitation	4,300,000	4,531,007
Safety and Medical	Not Split	2,350,270
Recruiting and Incentive Plan	see	7,294,589
Supplies, Light, Heat, Water, etc.	General	3,238,150
Compensation Insurance	4,150,000	392,493
F.O.A.B. and Unemployment Tax	4,900,000	5,904,486
General Item	11,564,279	4,215,069
Fees – Pipe and Electrical CPFF Contractors	— —	474,700
	$24,914,279	$28,400,764
FIELD SUPERVISION	12,860,752	8,963,101
PROCUREMENT, INSPECTION, WILMINGTON SUPERVISION	3,650,000	1,819,838
DESIGN ENGINEERING	1,500,000	3,280,062
TOTAL CAPITAL INVESTMENT	$300,000,000	$333,699,075
Value of Material transferred out by Government (Not included above)		28,941,496
Value of "Free Issue" Material by Government (Included above)		26,273,587
Value of Material purchased by Government (Included above)		27,622,404
TOTAL EXPENDITURES BY DU PONT COMPANY		$308,744,580

[a] On this date less than 13% of total drawings had been issued. (DuPont B – Design Status Charts)

Source: End Note

Consequently, the only feasible method of stating the equivalent 1994 cost, approximate and imperfect though it may be, is by cost indexing. It is necessary therefore, to examine the available indices, then select the one most applicable to Hanford.

The Engineering News Record (ENR) Construction Cost Index is customarily used for conventional types of construction, such as Hanford's infrastructure and structural components, but is not suitable for the process-equipment parts of the plant. This is because the ENR Index is composed of specific quantities of cement, lumber, steel, and common labor, and therefore says nothing about the costs of Hanford's process equipment.

The Hanford process components had a lot in common with chemical plants and refineries: stainless-steel piping and vessels; agitator vessels; centrifuges; steam and air-jet ejectors; an extensive, pressurized-water system; a very large water-treatment system; extensive normal and standby electric power systems; and large ventilation systems. The Chemical Engineering Plant and Nelson-Farrar Refinery indices are therefore candidates for Hanford's escalation.

A nuclear-power index might be applicable, but the earliest date of accurate cost input to such an index was 1966, according to Dr. James Hewlett, a consulting economist in DOE's Energy Administration. The reason is that the nuclear power plants prior to 1966 were built under turnkey, guaranteed-cost contracts, resulting in the turnkey engineer-constructors soaking up huge, unrecorded over-runs. The recorded costs of these plants are therefore not reliable measures of their true costs. Therefore a nuclear-power index beginning in 1966 would have to be back extrapolated 21 years to Hanford, and such a method would be an unacceptable flight of fancy. [2]

In table 7.2 I tabulate the values of the applicable indices, plus their increase factors relating 1945 to 1994. For escalation of Hanford's process components I decided to use the average increase factor – 10 – of the Chemical Engineering and Nelson-Farrar indices for the reasons of equipment similarity previously explained. For conventional Hanford components I used the ENR Construction-Cost index, with an increase factor of 17.7. For accounts that were primarily labor – fire and police, safety, medical, design engineering, etc. I used the Consumer Price Index, for which the increase factor is 8.2.

Table 7.2
COST INDICES AND MULTIPLIERS

	INDEX	1945	1994	Increase Factor
1.	Chemical Engineering Plant Index	58.9 [1]	366.7	6.2
2.	Nelson-Farrar Refinery Index	98. [2]	1337.8	13.7
3.	Average of Increase Factors for 1 & 2			10
4.	ENR Construction Index	307.8	5445.8	17.7
5.	Consumer Price Index	18.	148.2	8.2

Notes.

1. 1947 was the earliest year in which Chemical Engineering Magazine provided their plant index. I back extrapolated two years to obtain this synthetic 1945 value.
2. 1946 was the earliest year in which the Nelson-Farrar Index was provided. In a Nov. 10, 1993 telephone call, Gerald Farrar provided his estimate that the index would have been 98 in 1945.

The Escalated Costs

Applying the cost increase factors of Table 7.2 we have the following 1994 dollars (as calculated in Appendix G):
- Bare plant without first fuel charge $4.1 billion[a]
- Plant including fuel charge   4.4 billion
- Total Manhattan Project   27.0 billion

The escalated plant costs include the small profit permitted under today's DOE regulations – $18 million – or 0.44% of total cost of the bare plant. This amount is below the accuracy of the escalation, but I mention it to anticipate the reader's question. Addition of profit to the 1994 costs was necessary because in 1945 DuPont did the job for a one-dollar profit in order to avoid the charge of war profiteering that was brought against them in the Nye hearings of the 1930s.

This one-dollar fee in 1945 brought about a droll situation. When DuPont agreed to the contract, they stated their estimated completion schedule of three years. Actual completion required but two years.

In the final Federal audit the Government auditor reduced the dollar fee by 32 cents, in proportion to the reduction in construction time.

When I asked Matthias if the auditors were being funny he said, "I don't know, I think it was part of the contract." Groves said in his book that the auditors were serious.

In 1945 the Pasco Kiwanis Club heard of this deduction and wrote to Walter Carpenter, saying that the Kiwanis were proud to be situated near the Hanford project and enclosing their contribution to DuPont of thirty-two cents. The Kiwanis-Club letter is reproduced as Figure 7.1.

Mr. Carpenter, in his speech of acceptance of the Army-Navy "E" Award on October 9, 1945 remarked that he greatly appreciated the cooperation of the citizens of the surrounding communities. And then he mentioned the Kiwanis-Club letter as an example of the friendliness of DuPont's neighbors in the community. He then added that the letter was placed in the DuPont museum in Wilmington, as were the thirty-two cents. [3]

Manhattan Cost Relative to U.S. And German Economies

The United States could easily afford the Manhattan Project and the Germans could not. The American numbers are as follows:

Total Manhattan Cost $2.2 billion [4]
Total HEW Cost (Table 7.1) $0.33 billion

These costs are related to the wartime U.S. economy as shown next in Table 7.3.

---

[a] A more accurate escalation would apply the ENR multiplier to the costs of the conventional facilities in the process areas, but the 1945 costs of those conventional facilities are not available. The 1994 costs of those areas are thus understated by an unknown amount.

PASCO KIWANIS CLUB
Pasco, Washington

MEETS MONDAYS
12:10 P. M.
HOTEL PASCO

August 11, 1945.

Mr. Walter S. Carpenter, Jr., President,
E. I. DuPont de Nemours & Company,
Wilmington, Delaware.

Dear Mr. Carpenter:

At the last regular meeting of the Pasco Kiwanis Club a resolution was passed which reads as follows:

> "An article in a local newspaper states that the DuPont Company received only One Dollar profit from the operations at the Hanford plant and that an expense item of thirty-two cents was not allowed by an accountant, leaving a balance of sixty-eight cents. Thirty-Two members of this club are contributing one cent each to make up the difference and also placing their signatures to this letter."

We are very proud to be so closely situated to the Hanford project, and all of us feel very sincerely that we have had a part in this magnificent enterprise. We also hope that The Lord will see fit to direct the future efforts and achievements of this product into the right channel for the good of all mankind.

Sincerely yours,

Mel Swanson

Mel Swanson, President.

MS/lb
Enc. 32¢

[signatures of members]

Figure 7.1

**PASCO KIWANIS CLUB LETTER
TO WALTER S. CARPENTER**

Reference: End Note TBD

Table 7.3
Manhattan & HEW Costs
vs.
U.S. Gross National Product

	Billion Dollars		Days	
	Per Year	Per Day	Manhattan	HEW
Gross National Product, 1944 [5]	210	.574		
Days of 1944 Gross National Product to Fund Total Project			3.8	0.6

For added perspective, consider the magnitudes of the following work that paralleled the Manhattan Project: the United States put in place another $51.9 billion of public and private construction, which included $8.5 billion of military facilities additional to Manhattan. At this same time the United States also manufactured 8948 naval vessels, 88,410 tanks, and 324,750 airplanes. [6]

The United States economy was enormous!

Germany on the other hand, had an economy less than one quarter as large as the U.S. All of their resources were apparently required for the basics of conventional war; there was little surplus for advanced projects, and the decision was made to commit that surplus to the Peenemunde rockets and the jet plane.

Their first cyclotron was not operational until 1944, 12 years after Lawrence's. Albert Speer, the German minister of armaments during World War II, subsequently wrote as follows:

"Perhaps it would have proved possible to have the atom bomb ready for employment in 1945. But it would have meant mobilizing all our technical and financial resources to that end, as well as our scientific talent. It would have meant giving up all other projects, such as the development of the rocket weapons."

and

"At best, with extreme concentration of all of our resources, we could have had a German atom bomb by 1947, but certainly we could not beat the Americans, whose bomb was ready by August, 1945." [7]

DuPont's Management & Engineering Costs
vs.
ASCE Recommendations

The purpose of this section is to determine the legitimacy of DuPont's management and engineering (M&E) costs at Wilmington and Hanford. These M&E costs, abstracted from Table 7.1 and rounded to the nearest 1000, are as follows:

Hanford Field Supervision	$8,963,000.
Wilmington Procurement, Inspection and Supervision	1,820,000.
Design Engineering	3,280,000.
Total, Rounded	$14,100,000.

Two upward adjustments to these costs are necessary to provide equivalence with peacetime projects. The first is for wage increases, which were frozen during the war. The second is for profit, because DuPont earned only a one dollar fee, as previously explained.

Assuming wage increases equal to the CPI increases of 1943 and 1944, and a profit of 5% of the M&E costs, we have the following costs, percentages, and comparisons (as calculated in Appendix G).

Table 7.4

COMPARISON OF ADJUSTED M&E COSTS

	Adjusted Costs, Millions	Percent of Project Cost	% Recomm. By ASCE Manual 45 of date:
Total M&E	$15.7	4.7	5.2 (1968)
Design Engineering	$ 3.7	1.1	4 – 9 (1988)
Field Supervision	$10.	3.0	2.5 – 8 (1988)

These low percentages would have been acceptable for the most routine of industrial plants. Given Hanford's first-of-a-kind nature, it was a remarkable feat of engineering that DuPont was able to complete the HEW under schedule for these low percentages. [a]

Project Cost vs. Project Duration

It is axiomatic that a project quickly done is of lower cost than extended projects. The HEW literature however, emphasizes that "buying time with money" was one of the project's prime goals. Although money bought time at Hanford, it also bought money – lots of it, because operating expense at Hanford was very large, as explained next.

From Table 7.1 we have the following expense:

Camp Operation		$18,369,000
Field Expense	$28,401,000	
Less Recruiting & incentive Expense (See Matthias B,p.7 & Table 7.1)	-7,295,000	
Less CPFF Fees	-475,000	
Net Field Expense	$20,631,000	20,631,000
Field Supervision		8,963,000
Total Hanford Operating Expense (OE)		$47,963,000

The construction period – March 22, 1943 to February 25, 1945 – was 23.2 months, and OE averaged $2,067,000 per month. The peak rate, assuming triangular distribution, was twice that, or $4,134,000 per month.

Given an OE of this magnitude, the time-saving decisions – discussed next – resulted in large dollar savings.

---

[a] The ASCE no longer makes percentage recommendations. Instead they recommend that fees be negotiated on the basis of specific project conditions. The former recommendations quoted above, however, provide a good order-of-magnitude comparison.

Time-Saving Decisions

DuPont A itemizes 28 time-saving decisions and actions in the following categories: [8] re-design; ordering materials long in advance of design: buying from warehouse stock, rather than waiting for rolling; splitting of purchase orders among several vendors; warehousing of material for later delivery to vendors; improvement of manufacturing methods; and purchase of equipment for installation in vendors shops.

Of the 28 decisions, two are accurately quantifiable – the stainless-steel plate warehousing order (the Procurement section of Chapter 5) and the quintuple split of the biological-shield-block fabrication order. [9] I quantify these instances below.

Four other decisions can be estimated from data in DuPont A in combination with reasonable assumptions. These decisions are:
- Ordering pile graphite on a tonnage basis six months before release of first pile drawing
- Subletting the machining of 40 shield blocks from Read Machinery's total order
- Flame cutting of tube holes in the shield-block plates
- Installing Masonite sheets in stacks during shield-block assembly, rather than installing sheets one at a time.

The time for these four items was additive on the Critical Path and totaled an estimated 1.2 years. The cost saving due to this schedule improvement was:

1.2 years x 12 months x $2.067 mil./mo. = $29.8 mil.

The remaining 22 decisions were also the cause of shortening the HEW construction time but the time saved cannot be quantified from the available data, nor can we say exactly how many of these instances were in parallel with each other.

The Stainless-Steel Plate Storage Purchase Order

As described in the Procurement section of Chapter 5, DuPont paid $86,000 to have a Pittsburg shop store stainless-steel plate in a variety of thicknesses for subsequent shipment to pipe and vessel vendors, once pipe-and-vessel design had been completed. This saved three months of the separation-plant schedule, an event on the HEW critical path.

From the Building & Facilities Progress Charts of DuPont B the start of Cell-Building equipment installation was April 24, 1944 and the Bulk Reduction Building started on June 12. This was at the peak of construction, as shown on Figure 5.4; therefore the peak OE rate of $4.134 million applied. The three months of OE thus came to $12.4 million, for a benefit to cost ratio of 144 to 1.

The Quintuple Splitting
Of The Shield-Block Fabrication Order

DuPont awarded this five-way split order because of obvious scheduling problems. Taking Read Machinery as an example, we see from Figure 5.6 that Read required 189 days to deliver their first 97 blocks, or 1.95 days per block. Had Read been the sole supplier – which DuPont never contemplated – Read would have required 2428 days to complete the total of 1245 blocks. At the average OE cost of $2.067 million the total OE would have increased by $165 Million. No such thing was ever contemplated, but it illustrates the cost of delay and the benefit of splitting purchase orders.

Chapter 8

LABOR CONDITIONS

A number of labor-related conditions contributed to Hanford's rapid completion – the principal ones being a minimal number of work stoppages, competent craft workers, worker attitudes, and weekly hours worked. Craft turnover, recruiting, and safety were other interesting aspects of Hanford labor.

Lost time due to work stoppages amounted to 0.012% of total time worked (15,060 man hours out of 126,265,652).[1] Matthias confirmed this small amount in his interview – "Practically none" was his statement.[2] Stanton said, "During all the time I was at Hanford I was not conscious of any union problems.[3] None of the work stoppages delayed job completion; the longest of four total stoppages lasted 28 hours.[4]

There were four principal reasons for this good record: the Building Trades Council of Pasco agreement to set aside jurisdictional boundaries; their no strike pledge; Matthias' good relations with labor officials and the crafts; and the patriotism of the crafts, as explained in Chapter 10.

The reason for setting aside jurisdictional boundaries was that the Building Trades Council could not furnish sufficient craft workers, although their efforts to secure workers extended across the nation. Shortages existed in some crafts at Hanford, while others had surpluses. Under this agreement "the plumbers permitted boilermakers and machinists to work at plumbing and pipefitting work. Machinists found it necessary to let carpenters work as millwrights. This cooperation between various union craft representatives greatly improved the balance of crafts and made it possible to maintain working schedules and coordination of work to fit completion objectives," according to Matthias. "In fact", he said, "it is doubtful whether work would have been completed on schedule had these cooperative agreements not been reached."[5]

Frank Matthias assiduously cultivated the stewards, business agents, Presidents of labor groups, and officials of the Internationals. Approximately one quarter of his diary entries concerned meetings with these officials. He remarked, "I did talk to them a lot, and my door was always open to the business agents and stewards."[6]

He seems to have gotten along with everyone, even Dave Beck. Beck was President of the Western Conference of Teamsters, headquartered in Seattle, and according to the post war newspapers, was an extremely contentious individual. Matthias remarked however, "Dave Beck was a very cooperative guy. Whenever we decided in a meeting that he and I were going to do things in a certain way, then he always did the very best he could to do what he promised."

"He ran this little newsletter for the teamsters under his jurisdiction, and I'd read it and once in a while I'd find something in it that was not right and I'd phone him and tell him, and he'd say, 'Well, write it the way you think it is.' And I would and I'd mail it to him, and he would always print it in the next newsletter."[7]

Getting along with labor, however, went only so far with Matthias. In July of 1944 the pipefitters had a jurisdictional walkout and had a meeting in the recreation hall instead of reporting to work. Matthias had the following account of that meeting:

"I took the stage in the recreation building – it held a thousand people – and they had loudspeakers to talk to themselves. Apparently they had a date with a union guy – an official – to be there to help them. There was one guy on the stage that I knew – a young fellow that was a pipefitter, and I said, 'Give me a chance at that microphone.' And he said, 'Good, these guys sure need something.' "

"So I gave them a little talk, and I said, 'You know, you guys are breaking your promise. Your union agreed that we'd have no strikes. Now far more important than that, you are having a strike and not working on a project that is of tremendous importance to this country. Now, I thought you appreciated that, but apparently there are some of you that are moving this strike, and they're working for the Germans, and if I knew who it was that was fomenting this thing, I'd arrange to have them sent back to Germany, where they belong.' "

When I asked him how they took that, he said,"If they had guns they would have shot me! And after I got them quieted down, I said, 'Look, take it easy. I'm not calling you traitors, but some of you are acting like it. Now how about going back to work and doing what you promised, and what we need badly. I'll have the buses at the door in ten minutes.' And then they cheered. And that was the end of it. They all went back to work, and their union leaders came as they were rushing out the doors, and they wouldn't go back in to listen to them. And I said, 'I'll set up a deal, and we'll discuss all these problems you have and we'll do something about it, and you set up a committee, but I don't want a committee that comes from guys that are hardly working. We need to be working.' We settled our discussions with the pipefitters the next afternoon."[8]

Contrary to some accounts of Hanford, the construction crafts knew their job. Matthias said, "I think they were pretty well skilled if they said they were. I never saw any unskilled workers."[9] Stanton agreed:"All of the standard crafts knew what they were doing when they came to Hanford."[10]

I asked Mr. Stanton about the problem of finding highly qualified weldors for the specialized biological shield block welding. "I never had any shortage of qualified weldors," he said. "We'd get 'em in and give them two or three days to show them what we wanted, and they did fine."[11]

In fact, it was basic War Department recruiting policy that the recruiters screen out sub-marginal workers during the interview period.[12]

Because the crafts were competent when hired, DuPont did not have a major crafts training program, as other accounts had it. "We didn't train them," said Matthias.[13]

A final observation on craft competence was the 1993 statement of a member of DOE – Richland's Site Infrastructure Group that whenever he visits the old 105 Areas, how impressed he is with the quality of the workmanship.

Stanton reported that the crafts in the reactor buildings were interested in what they were doing. "Everything was so different from what everyone had done before," he said. "All the crafts came up with valuable suggestions – things that had not occurred to the engineers." He thought that "the mystery of it all had a positive effect. It was intriguing – it piqued their interest."[14]

This reaction was confirmed by Parker's observation when he was transferred to the reactor building after working in areas of conventional construction. "The first time I saw that atomic pile", he said, "I was almost staggered by the spectacle. It was immense. I spent quite a few free moments contemplating this structure, but I knew no more than before I entered."[15]

Secret Weapons

From the cover of the men's employment-information brochure

ABOVE: HANFORD SCIENCE MUSEUM
BELOW: LAWRENCE BERKELEY LABORATORY/BATTELLE PRESS/THE PLUTONIUM STORY

ABOVE: The employment staff poses for its group portrait with new craft workers.

LEFT: One of the Hanford mess halls.

Stanton cited the instance of a last-minute Met Lab addition of a vent valve to each of the 2004 fuel tubes, which was completed in three days of forced draft effort. I asked if the pipefitters resented being pushed extra hard. "Not at all," he said. "They understood it was an emergency and wanted it done as badly as we did."[16]

The basic work week was two shifts of six nines each, with an hour between shifts. Shifts additional to the basic two were occasionally worked when the job fell behind.[17]

One of the reasons for Hanford's good construction productivity was Sunday work. As Walter Simon remarked, "We worked enough people overtime on Sunday so that what the rest of them had to do on Monday got off to a good start. They used the day off to catch up on a lot of little things that would be in the way next week. They were buying time. The Army said, "Don't hesitate to use money to buy time. Every day may be significant."[18]

There were 94,307 hires to maintain an average force of approximately 22,500, giving a turnover ratio of 4.2:1.[19] This turnover rate, although a challenge to recruiters, did not significantly impede progress. With DuPont's engineers providing precise daily instructions to the crafts, followed by hour-to-hour supervision, it didn't matter much which individual craftsman was involved, so long as he was competent.

DuPont's recruiters, numbering 153 at the peak, fanned out across 47 states, Alaska, and Canada and interviewed a total of 262,040 applicants.[20] This effort was initiated, with the participation of the War Manpower Commission and the U. S. Employment Service, at the beginning of the project because it was then recognized that the State of Washington provided an insufficient worker pool, given the war projects then in progress in western Washington.[21]

The construction force curve is included as Figure 8.1 and is taken from Matthias B, p. 55. (The peak force of 44,000 shown on the curve is contradicted by all other sources, which give 45,000. I have used the latter number throughout the book.)

Hanford apparently had an effective safety program. A Major Injury Chart for Hanford, dated February 18, 1945 when the project was complete shows 613 total major injuries.[22]

Dividing this into HEW's 126,000,000 man hours gives 206,000 man hours per major injury. This chart also notes that the corresponding rate for the general construction industry at that time was only 0.379 times as safe as Hanford, or 78,000 man hours per major injury.

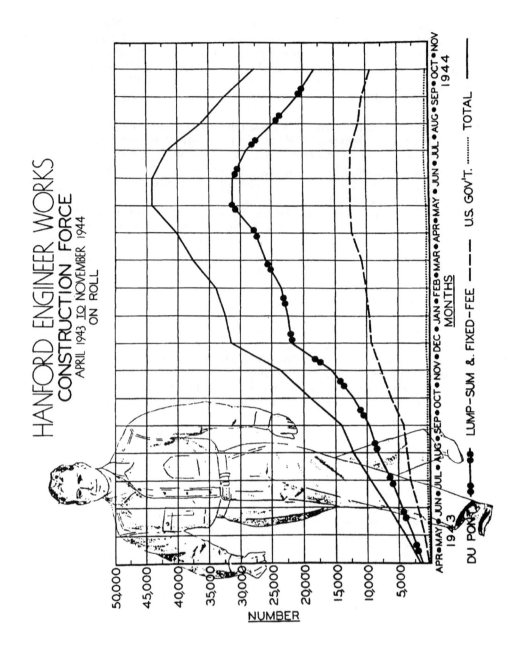

Figure 8.1

HANFORD ENGINEER WORKS — CONSTRUCTION FORCE

Reference: End Note 2

Chapter 9

TECHNOLOGIES OF THE FORTIES

The rapid completion of Hanford and the other large projects of that time should be put in context with the technologies of a half century ago. Some of these obsolete technologies are described next.

Construction

Cable scrapers, bulldozers, and cranes were used instead of today's hydraulic rigs. The scrapers worked from slope stakes, rather than from automatic laser guidance. The operator would occasionally get down off the tractor to check the slope stake numbers until he had the grade fixed in his mind. The D 8 was the largest tractor available at that time and had only 132 HP (98 kW). Today's D 11 is 770 HP (574 kW), and its bulldozer blade at 45 cu yd[1] (34.4 cu m) has almost twice the earth moving capacity as the largest *scraper* available to us in the Pacific War in 1944! Sheepsfoot rollers were tractor towed then; the self powered unit had not yet been developed.

Today's very large scoop loaders did not exist; instead, the less maneuverable, smaller capacity clamshells and power shovels were used for basement excavations and for loading trucks from embankment faces or from material stockpiles.

Gantries and time consuming guy derricks were used. Many crawler and truck cranes were required at a work site because tower cranes were not used in the U. S. Concrete was often placed by crane hoisted bucket and by buggies on temporary and time consuming catwalks.

Plastic underground pipe didn't exist. Time consuming cast iron and welded steel were used for smaller diameters of water and gas mains.

Surveying

Surveying was done by the same transit, chain, plumb bob, Philadelphia rod methods that built the late nineteenth century railroads and the Panama Canal, and was recorded in the field in hand lettered notebooks as input to subsequent hand calculations and drafting. Searles and Ives in their 1945 edition cautioned us that "cross sections should be traced in ink at the first opportunity to secure their permanence." The same edition also admonished us to record our culvert layouts in our "masonry book." Trench and pipe invert alignment was by transit and rod; the automatic, laser target system was in the future.

Process Instrumentation

All sensors in those decades reported to panel boards in the central control room. On these boards were mounted large analog instruments, 10 to 12 in. (25 to 30 cm) in diameter. (The miniature analogs came in the fifties.) The boards for a large process plant were several hundred square feet (several tens of square meters) in area. Pneumatic instruments reported via literally miles of copper tubing run in overhead racks.

Today's distributed control, reporting via coaxial data highway to central PC monitors, is far quicker, and less expensive to install than was the instrumentation of the forties.

Engineering Methods

Slide rules, hand cranked calculators, and numerous charts, tables, and nomographs took the place of computers and hand held microchip calculators. Everything therefore was hand calculated, even such tedious work as Hardy Cross nets. Precise, seven place calculations, like traverse closures, required hand interpolation and transcription from logarithmic and trigonometric tables, followed by hand or hand cranked calculator arithmetic. Graphical solutions and presentations required hand plotting on printed arithmetic, logarithmic, and calendar graph paper. Such graph paper is nearly unobtainable today, having been superseded by computer-generated graphs.

Drafting machines were often available but T squares and triangles were still in use. Reproduction of opaque originals was avoided because that meant costly photostats at an outside firm, many blocks and 24 hours away. Abstracts, or complete copies were typed from the opaque original for limited distribution if the topic were urgent. If the opaque were thin enough, one turned up the ozalid intensity and got a faint copy, or rendered the original transparent by oiling.

Ozalid machines, however, were sometimes an unobtainable luxury. In June of 1941, having retrieved a tracing from the files of an office in Ukiah, California, I asked directions to the print machine. With a sly grin the office engineer displayed his sun frame, in which I placed the blank paper and the tracing - then stood outside for 20 minutes or so on the main street of Ukiah until my print developed. On rainy days, I presumed, prints were not made in Ukiah.

Neither Configuration Management, Systems Engineering (as opposed to the engineering of systems), nor data trees had been thought of. ANSI/ASME NQA-1 was then 35 years in the future.

Office Methods

Large, multi page spreadsheets, such as the monthly cost and comparison to estimate report, were handwritten on printed ledger paper using the hand cranked calculator and then typed on long carriage typewriters. Because of inevitable errors, the final row and column totals often differed, requiring time consuming error searches.

Typewriters were manual; corrections were by eraser – tedious with five carbons. Planned reproduction was by mimeograph, multilith, ditto, and carbon paper, all cumbersome, usually filthy, and always time consuming, Ozalid opaque transparencies were typed but this work was slow.

Control of procurement, inventory, and equipment accounts was by hand on printed matrices on cards or paper, on which the daily changes were entered by hand from purchase orders, invoices, warehouse receipts, equipment use tickets, expediting reports, and so forth. Monthly status then had to be hand totalled – then transcribed by hand to the draft reports for subsequent typing.

Figures 5.6 through 5.9 are examples of manual procurement and inventory control sheets used on the Hanford project.

Filing, file indexing, and file retrieval were hard copy, by hand. Hanford's facilities inventory alone occupied 35 filing cabinets full of file cards.[2]

Transportation

If everything went just right, it took 18 hours to fly to Pendleton, Oregon from Washington, D.C., and that was followed by a two hour drive to the Hanford camp from Pendleton.

The first site investigation trip however, was an example of frequent delays. The site team was grounded in Chicago for 24 hours, and for five hours in Cheyenne, resulting in a total elapsed "flying" time to Seattle from Washington, D.C. of 48 hours. On a trip from Washington, D.C. to Pasco, Matthias flew via Chicago to Minneapolis, but was grounded at Bismarck, where he took the train; elapsed time was 67 hours.

The Manhattan literature mentions the Twentieth Century Limited, a high speed, steam driven, luxury express which left Chicago in the afternoon and arrived in New York City at 8:00 AM the next day. Transcontinental trains required four days for the coast to coast trip. Simon though, preferred the train to air travel. "I took the train because once I got on the train they weren't going to put me off. But I could be bumped at any stop on an airplane."[3]

There were no long distance, four lane, divided interstates – only the interminable and twisting two lane highways (and the killer three laners), with frequent local access intersections and roadside businesses, all routed through every city, town, and hamlet, and all with frequent stops at train crossings and city stop lights.

The U. S. Mail

First class mail was inexpensive but as slow as the railways. Coast to coast and sometimes San Francisco to Los Angeles took four days. Air mail was one and one half to two days, but expensive.

Long Distance Telephone

These calls were placed by operators sitting at plug in switchboards. She disconnected the caller while she tried to find a routing through the national network. She would reconnect in a few minutes on a quiet day, but one hour delays were not uncommon.[4] For very important calls, one sometimes sent prior telegrams to be sure that people at the far end were in place and waiting.

Long distance was very expensive from one's personal point of view and therefore usually reserved for emergencies. The operator's announcement, "Long distance calling" was therefore dread cause for alarm.

In business, the long distance call was reserved for the very unusual case, at least in engineering offices. Seaborg and his associates, for example, almost always wrote letters for arranging meetings and other routine matters. In the three months ending June 15, 1942, for example, Seaborg's diary recorded 22 such letters.[5] Every one of those instances would have been telephoned today, given the present, low cost, digital switching.

Chapter 10

INTANGIBLES

It is impossible, by definition, to quantify the effects of most of Hanford's intangibles on schedule and costs. It is certain, however, that the principal intangibles described in this chapter, played a significant part in the completion of Hanford a year ahead of schedule and for a relatively small overrun.

The management forces at the Met Lab, Wilmington, Hanford, and in Washington, D.C. were strongly motivated. They voluntarily worked 10 to 20-hour days, most of them seven days a week. Genereaux said that he worked so many hours he dreamed about it. [1]

Simon and Matthias were responsible for very large projects during their post-Hanford careers but they both found Hanford the most exciting job they worked on, before or after. Simon remarked, "After Hanford everything else was an anticlimax." Matthias mentioned "the tension and the challenge" of Hanford, "and the fear of Germany's doing it first." [2]

Fear of Germany strongly motivated all of those in high management because they knew the secret of the bomb. Compton and Wigner in June of 1942 calculated that if Germany knew about plutonium, they might have six bombs by the end of 1942. The Met Lab then discussed infiltrating their physicists into Germany to destroy their reactors! [3] A Seaborg diary entry reads: "The Germans may have beaten us to it. I wonder, are they aware that ^{233}U can be made from ^{232}Th, and 239_{94} from ^{238}U in a chain-reacting pile and that either of these isotopes can be used in a fission bomb?" [4]

Stanton said, "There was a definite incentive in the national emergency." Simon pointed out that "Almost everyone had relatives in the service, and every week someone in their family or that that knew was killed, so everyone felt very keenly the necessity of what they are doing." [5]

The voluntary cooperation of industry was outstanding, as the following examples illustrate. Compton went down to St. Louis in 1942 to talk Edward Malinckrodt into immediately purifying sixty tons of uranium by the ether process. Malinckrodt went ahead without any formalities, on Compton's assurance that someday the OSRD would come up with the proper paperwork. Malinckrodt shipped the last of the order the day before the contract was signed. [6]

Industry also cooperated in security during preliminary conversations regarding possible contracts. As Groves pointed out, "The urgency of the project did not allow time for us to conduct any detailed security checks in advance of negotiations; instead, we relied upon the discretion and patriotism of American industry. We considered this a good risk and we were never disappointed." [7] Those familiar with the time-consuming nature of security checking will recognize that this cooperation on the part of industry saved many months, even years.

And, as Matthias pointed out, "This whole thing was put together with such faith people had in others – the leaders – it was just amazing!"

"Did I ever tell you when Groves proposed this project – that DuPont take it over?" I said I hadn't heard. "Alright, he went to Wilmington and talked to all the hot shots. And he finally had to tell them what it was all about. And he went to a meeting in Wilmington the day they accepted the project, after he made his pitch. There were about ten or twelve people at the table – the top of DuPont and Groves. And he had a folder with a lot of information on it at each chair. He opened the meeting by saying, 'Now what's in front of you is a discussion of what this project is for. It's very, very critical information that you people should know, and we've talked about some things that are beyond our normal security limits. This goes into a lot of detail, and I'd rather you didn't read it, but you're at liberty if you have any doubts or questions at all, you can read it.'"

"And there wasn't a single DuPont person in that whole group that touched that thing until they promised Groves they'd take it." [8]

A final point on industry cooperation illustrates not only industry's attitude, but the prevailing outlook of all of us. I had quoted to Lombard Squires Nichols' statements in Chapter 4 about the total lack of financial irregularities in the Manhattan program. Squires remarked, "We all wanted to help the war effort." [9]

Among the academics of the nation, after Pearl Harbor, the only question was which project needed them the most. Almost every scientist that went to work on war projects set aside indefinitely the personal research that was central in his professional career, but they willingly did it because of the importance of the war effort to the nation. [10]

The Hanford crafts were solidly patriotic. On their own initiative they asked everyone for a day's pay and raised $162,000 in seven weeks, enough to buy a four-engine B-17 bomber for the Air Corps. The bomber, named the "Day's Pay," flew in from Boeing Field in Seattle to the Hanford air strip in July of 1944 for a memorable ceremony to present the bomber to the Fourth Air Force. Matthias remarked, "The principal talk at the ceremony was by Major Grashin, one of the officers who made the 'Death March' on Bataan. His talk was inspiring and he spared his audience none in his descriptions." Concerning the overall effect of buying the B-17, Matthias said, "This activity, conceived by the workmen and handled by them, coupled with the talk by Major Grashin was the most effective single morale builder during the job and did much to develop an attitude of teamwork and desire to help the war than any other thing." [11]

There were many negative intangibles, of course. Off-the-job boredom was always in the background. Town was 40 miles away, and you couldn't get there very often because of gas rationing. And, as Simon remarked, "Richland was a very sparse village – no bakery or beauty parlor, one drugstore, one food store, one clothing store – people waited in line for everything. There was only one barber shop; they allotted six minutes for a haircut. They didn't want to bring a lot of barbers in – it would just take up housing they needed for production people." [12]

Parker reported that after a ten-hour day and an hour commute each way, sleeping was the only activity that amounted to much. As far as the camp movies were concerned, he said that the lines from the ticket office were four hours long, so you had to be a pretty dedicated movie fan to take part in that. [13]

Welch mentioned that many of the men in the barracks were rough guys – a lot looked like ex cons. There was gambling in the barracks and lots of fights. [14] The police had now and then to break up the beer-hall fights with tear gas. The camp appears to have been the civilian version of the Army cantonments experienced by millions of us during the same period Hanford was in progress.

BOTH PHOTOS: HANFORD SCIENCE MUSEUM

ABOVE: The memorable "D[ay's] Pay" ceremonies in which th[e B-]17 G bomber, bought and p[aid] for by Hanford's craft work[ers,] was transferred to the Fourth [Air] Force.

LEFT: Central administrat[ive] area of Richland. Housing [in] background. Transient quart[ers] just out of picture, right cen[ter.] Large, multi-winged building [in] center is the Federal Buildi[ng,] since replaced by the curr[ent] seven-story building on the sa[me] site.

100

BOTH PHOTOS: HANFORD SCIENCE MUSEUM

ABOVE: Hanford camp, too large for a single photo! Trailer camp in foreground; barracks in background. Administrative area in center, rear. Columbia River, the White Bluffs, and Franklin County as backdrop. Note the very light traffic. Walter Simon remarked, "You know, two gallons of gasoline— that's all we got a week."

LEFT: The camp heating plant. Note the steam source— five locomotives. Wood stave tanks were used to conserve steel.

Richland's Transient Quarters

This building became the post-war, privately-owned hotel, the "Desert Inn," subsequently torn down and replaced by the current "Hanford House" in the same location.

HANFORD SCIENCE MUSEUM

Welch also remarked that, although the men didn't much like Hanford, it was just after the Depression and they were making more money than they ever had, so that helped to keep them on the job in spite of Hanford's disadvantages.

The dust, the monotony – all the camp buildings were painted "Government Green," according to Parker – the lack of much to do make it remarkable that the morale of those who hung on remained relatively high. It was a tribute to the workmanlike attitudes and patriotism of the craft workers.

Chapter 11

WHY HANFORD SUCCEEDED

A Summary Of Lessons Learned

Hanford came in with a near-perfect start a year ahead of schedule because of the intangibles – summarized at the end of this chapter – and because of the use by the Corps and DuPont of a large number of methods and attributes, all of which had been deliberately adopted, either specifically for the Hanford project, or previously as part of their standard procedures.

This chapter is a summary of these methods and attributes – collectively designated for simplicity as "methods" – all of which have previously been described. Beyond applicability to Hanford, most of them are also necessary for any project considered so important that it cannot be allowed to fail, and which is also conducted under Hanford's conditions of fear and urgency. In that sense these methods are in the nature of general lessons learned.

Many of these methods were common to other large projects of that time and thus were not unique to Hanford. Some, such as managing by a single authority, are common sense and should be commonplace, but often are not. Some, such as giving a single federal agency sole jurisdiction over site selection, would violate current laws.

The exact ranking in relative importance of these methods is open to debate, but not debatable was the necessity for each of them. The absence of any of them would have delayed delivery of the final product by many weeks. The absence of many of them would have delayed completion by years.

It is, of course, understood that a principal cause of Hanford's success was the early start made by the Met Lab, and their intense efforts in the thorough definition of the physical and chemical concepts. This chapter, however, is a summary of the subject of this book – engineering and construction management.

The following summary is divided into the General Lessons Learned and the Specific Methods and Practices. Together, these two categories added up to success.

General Lessons Learned

1. The economy of the U. S. was and still is so enormous that very large emergency facilities of almost any magnitude and cost may be planned and executed with confidence.
2. Grave uncertainties and unprecedented difficulties need not be insurmountable barriers to rapid project completion. These uncertainties and difficulties can quickly be overcome by method, forethought, imagination, ingenuity,

determination, and organization, especially when motivated by urgency and fear.

3. In a matter of survival of the nation it is essential that national leaders exist who are capable of mobilizing the national will.
4. It is imperative that national leaders have the discernment to distinguish between the nation's interest and their own, and the integrity and fortitude to act on the basis of the former.
5. Total program authority must be vested in one executive in one federal agency. This means that parallel federal agencies that would normally participate will do so only on a cooperative, advisory, and assisting basis. It also means that these parallel agencies will have neither the authority nor the opportunity to interfere in decision making, and will be instructed that the utmost in cooperation is required of them.
6. The program executive must have complete support from the highest level, the President in the case of a national program.
7. Weak points in the organization must be identified early and the necessary steps taken to correct them.
8. The program organization should be simple and direct and the command channels must be clearly understood by all management levels.
9. Authority and responsibility must be kept together.
10. Large staffs are to be minimized or eliminated; they are a source of inaction and delay.
11. Decisions should be made on the spot as they are needed.
12. Written communications among the top managers should be minimized or eliminated. Face to face or telephone conversations should be used instead.
13. The managing agency must be given the top priority for all material and personnel resources.
14. Defense Acquisition Policy and procedures must be reduced to the bare minimum, meaning that procurement approval will be assigned to one or two high level managers. These managers must therefore be of demonstrated capability and integrity.
15. Crash engineering and construction programs of this type must be managed by a turnkey organization. The most experienced possible turnkey engineer constructor must be chosen.
16. It will be necessary in this type of crash, science-based program to proceed in parallel with the development of the supporting science, conceptual and final design, procurement, the semiworks, and construction – then exert the required and very considerable effort to ensure that all of these phases converge successfully at project completion.
17. The immediate design of facilities needed for early construction means that the design of these facilities must be frozen early in the program. This in turn means that these frozen designs must be flexible enough to accommodate any of the then-undeveloped alternative process concepts still under consideration.
18. If the alternative process most likely to succeed cannot be identified in a list of several candidate processes, then build all of them. Money under these circumstances is less important than workable results.
19. For financial control an auditing organization should be established at each project location that (1) is constantly available to advise procurement and

accounting personnel, and (2) will complete the audit of every expenditure within 30 days.

20. A positive expediting procedure should be instituted that puts experienced expediters in the vendors' shops to anticipate potential delays in time for the vendors to do something about them.
21. The type of program described above will require highly qualified managers in both the federal agency and the turnkey engineer contractor. They must be broad gauge, experienced people who can work together on the basis of mutual trust, and who have demonstrated in their previous careers that they have the following characteristics:
    - Dedication to cooperative working relationships
    - Superior management capabilities
    - Superior judgment
    - Superior decision making capability
    - In the case of engineering managers, exceptional technological experience and skills

They must be able to work without large staff organizations and without confirming documentation or written instructions. This means that they have the mental capacity to carry a myriad of requirements and agreements in their heads.

Specific Methods And Practices

1. In the engineering of untried and unprecedented concepts and components, exercise extreme engineering conservatism, plus painstaking care and caution.
2. Pay total and constant attention to detail.
3. Care, caution, and detail take precedence over schedule.
4. In producing drawings, detail everything. Leave nothing to "common practices of the crafts."
5. Produce operationally realistic drawings by having operations people work with engineers.
6. Eschew elaboration in design.
7. Check 100% of engineering and drafting by means of a checking group independent of the design group.
8. Hanford had the then-unique problems of remote maintenance and remote operation of the process equipment in the separation plant. This meant the following:
    - Equipment was designed to very close tolerances – hundreds and thousandths of an inch. This required extra care in drawing production and in vendor supervision.
    - DuPont made extensive use of their experimental and developmental Wilmington Shops.
    - The design was verified by fabricating components in accordance with preliminary issues of the drawings – then installing the components in mockups in the Wilmington shops to identify design deficiencies before sending the drawings to fabricators and to the field.
    - Each operational assembly was assembled in a mockup at Hanford before actual installation in the hot cells to verify/modify dimensions of all components.

- All assemblies required to operate in a remote area were installed by use of the remote maintenance equipment, thus ensuring in advance that the remote maintenance equipment and their operators were indeed capable of the future remote maintenance.
- All of the unique and original handling equipment was manufactured in the Wilmington Shops.

9. Procurement of Hanford's exotic equipment on a very tight schedule required the adoption of a number of unusual procurement methods, as follows:
   - Reserving space in vendors schedules by early orders based on quantity estimates made prior to design.
   - Waiving competitive bidding if that would save time.
   - Placing verbal orders concurrently with requesting Corps approval. The Corps guaranteed compensation to DuPont for cancellation costs if the Corps subsequently disapproved one of these verbal orders.
   - Splitting large orders among several vendors despite differing bid prices, if splitting would save time.
   - Buying from warehouse stock, rather than waiting for the less expensive mill rolling, whenever the schedule was in jeopardy.
   - Procurement of used equipment to avoid the time required for procurement and fabrication of new equipment.
   - Buying machine tools for loan to vendors to increase their production rates.
   - Permitting lower tier orders to subsidiary vendors.
   - For procurement of special, exotic equipment, establishing ad hoc control/engineering/expediting groups.

10. Hanford scheduling was by the Critical Path Method, which DuPont had originated about 1940. For the subsequent 19 years DuPont was the only construction company to use it.

11. DuPont managed all of their jobs, including Hanford by the engineer supervision method, in which the Assistant Division Engineers issued written instructions to the craft foremen for the next one to three days' tasks, and then constantly monitored the job, correcting the foremen's activities, as necessary.

12. DuPont conceived and used a Quality Assurance method for the reactors and the separation plants entirely the equivalent of ANSI/ASME NQA – 1, anticipating the requirements of that document 35 years before its publication.

13. The OIC, the Building Trades Council of Pasco, and DuPont agreed to a no strike pledge for the duration of the job and to set aside jurisdictional boundaries if sufficient manpower was unavailable in any craft, as it was frequently not available.

    The patriotism of the crafts and Matthias' painstakingly-cultivated good relationships with labor officials at all levels were major causes of these relaxations of normal labor practices.

14. All of Hanford's craft workers were competent; they weren't hired if they were marginal, nor were they trained at Hanford.

15. The work week was 108 hours – two shifts of six nines – plus Sunday work and other overtime shifts as required.

16. All of the managers, at least down to the fifth management level, worked between 10 and 20 hour days, seven days a week for the duration of the job. This was true at Hanford and Wilmington and for most of those at the Met Lab.

Principal Intangibles

These were:
- Strong motivation of management at Hanford, Wilmington, and the Met Lab, as well as Washington, D.C.
- Fear of Germany among those who knew the secret
- For all, the incentive of the national emergency
- Keen awareness of the personal impact of the war on the families of management and crafts alike
- The voluntary cooperation of industry and academia

Chapter 12

AFTERWORD

Hanford was an extraordinary management achievement. Despite the daunting uncertainties, the incomplete status of the new physics and chemistry, the unprecedented engineering and manufacturing difficulties and the distinct possibility of total failure Hanford was completed successfully a year ahead of schedule and an astonishingly small 11% over budget.

It may be thought however, that this entire book has been only the account of a quaint period piece – we would never do it that way today. We now have far superior methods in engineering and construction to those of a half century ago. True enough, but having said that, how would we do it today? We cannot escape the fact that back in those antiquated days and with those clumsy methods this remarkable thing happened. The fact of this historic accomplishment and the way it was done are the story of this book.

As pointed out in the last chapter, it was done by the application of thorough, well-thought-out management methods. Forethought, ingenuity, and method compensated for the absence of precedent.

Compton called the plutonium project "a miracle of coordinated effort," and the atomic program "a heroic act of faith." From the perspective of a half century, his words ring truer than ever.

Was Hanford unique? In the sense of rapid completion of a scientifically and technologically original plant, under the necessity of the gamble on the parallel-activity method, yes, Hanford and its companion plant at Oak Ridge were not only unique, their like has never been repeated.

It is necessary however to put the Hanford achievement into perspective – to relate that project to the circumstances of the time, particularly the other great projects of the Thirties and the Forties. In quick construction alone Hanford was not unique.

Everything was built quickly then. DuPont's rapid contruction of the pre-Hanford war plants has previously been mentioned. Kaiser's first cement plant, carved out of California's precipitous Coast Range, required but nine months in 1939. His Richmond, California shipyards in 1941 required nine months from bare mud flat to first launching. An additional yard went from ground breaking to first launch in four months.

The Kaiser efforts, like Hanford, were accomplished by well-thought-out management methods. The Richmond yards, like Hanford, were built with turnkey management. And, like Groves' executives and DuPont's, Kaiser's men communicated mainly in person and by telephone. Both DuPont and Kaiser were moving so fast then that by the time a memo could have been delivered the associated task should long since have been done.

Those still alive who managed those other quick jobs for those such as Kaiser, Atkinson, Walsh, Stone & Webster, J. A. Jones, and all of the other notable builders of that time may read this book and say, "So what? We did the same thing."

Were the management methods used by those other old-timers the same as for Hanford? I imagine they may have been similar but, except for the above-mentioned similarities, I don't know. Maybe someone should talk to these survivors, and document some of the other great projects of that era, but quickly! Time is fast running out.

A Dire Future?

As mentioned earlier, Hanford was built under the spur of fear and a sense of urgency. In some future year, distant or near, when the Nation may again in a matter of a mega-project on the outer fringe of science experience fear and urgency, might it be that the methods that built Hanford would again prove useful?

Two examples of such a future emergency come to mind: the hydrogen-fluoride (HF) laser satellite and the trans-atmospheric vehicle (TAV).

In a November, 1993 telephone conversation with a deputy manager of the Pentagon's Ballistic Missile Defense Organization, I explained some of Hanford's defense-acquisition methods. The deputy immediately understood the implications for the acquisition of the HF satellite, an exceedingly useful weapon (although not yet funded) for destroying enemy missiles in the boost phase. Although Hanford's experience was new to him, he had the perception to understand that the Hanford methods, if applied to specific aspects of HF development might reduce the acquisition cycle from 15 years to two and one half years. "It all depends on the national will," he said. [a]

The second example, the TAV, was conceived by studies done under the auspices of the Air Force and is now under preconceptual study by the Phillips Laboratory's Propulsion Directorate at Edwards Air Force Base. The concept is for an "airplane" that can take off from land and perform successive missions in space and in the atmosphere, alternating between the two as required. But how can this be possible, given the huge and clumsy auxiliary fuel tanks required to get the space shuttle up just once?

One Air Force concept is the use of *anti-matter* as the fuel! According to a conversation with the Propulsion Directorate, they have conceptually identified the magnetic bottle — like that at Princeton's Tokomak – as the "gas tank." The fuel "refinery" was to have been the Superconducting Supercollider, which of course has been cancelled by Congress.

Now relating all this to Hanford, visualize the panic in some future year following a CIA briefing of the President to the effect that Nation X is certain to have their TAV deployed in six years, perhaps four on a crash basis. Ours however, is still only in the pre-conceptual stage and our Superconducting Supercollider is only a set of drawings in an archive. The land on which it was to have been built has long since been sold, so the siting process must begin anew.

You will then have an exact parallel with Hanford's program – another mega project under the spur of fear and a sense of urgency on the far edge of science, and once again Hanford's methods will be necessary.

---

[a] It is true, of course, that there are today, as in this HF example, people as smart and perceptive as the Manhattan team was. A recent example was the crash re-design/re-building crisis at Milliken Carpet in Georgia in six months of dedicated, 24-hour days and seven-day weeks by Fluor-Daniels and DuPont after a January, 1995 fire had destroyed this 700,000 sq ft plant (Engineering News Record, Sept. 18, 1995, p.28).

POSSIBLE ADDITIONS TO THE RECORD

Among the interesting, but so-far-unavailable additions to the record would be:

- Prints of some of DuPont's earliest CPM drawings, showing the development stages leading to the final method
- Some of DuPont's in-house correspondence concerning CPM development
- The Met-Lab flow sheets and letters available to DuPont on October 3, 1942 at the start of separation-plant design
- The complete set of work orders from TNX to the Wilmington Shops, which would quantify the tremendous role played by the Shops in Hanford's development
- The complete set of Stanton's 3000-plus pages of QA procedures for reactor construction, and the similar documentation for the separation plant
- The handwritten, Matthias-Church-Hall site-trip report of Dec. 31, 1942
- A then-classified movie of Hanford construction made during the war. Col. Matthias remembers seeing the Signal Corps making the movie, but queries to various Government agencies have not yet disclosed its existence.

On a lower level of detail, but still of interest would be:

- A force curve for the TNX Design Group, and a split of the hours devoted to engineering of the Clinton semi-works. These data, combined with the available Design-Status charts would permit calculation of average hours per drawing.
- Total costs of the development purchase orders charged to TNX Design
- Hours spent by Design at Hanford and in monitoring development purchase orders
- Separate costs of process and conventional facilities in the reactor and separation plant areas (100 and 200) to permit more-accurate cost escalation to 1994

Chapter 13

ACKNOWLEDGEMENTS

Completion of this book required a lot of help from some very cooperative people, at no small cost of their own time.

There would have been nothing to write about without the first-hand accounts from the original scientists, engineers, and craft workers – in chronological order: Col. Franklin Matthias; Professor Glenn Seaborg; Walter Simon; Raymond Genereaux; John Tepe; Russell Stanton, Lombard Squires; James Parker; William Welch, and Robert Nelson.

Genereaux reviewed the complete manuscript and Stanton the section on Hanford management. Stanton also searched his records to provide the details and chronology of the origin of CPM. Squires briefed me on the relationships among the Met Lab, TNX, and the Engineering Division, and also pointed me to Seaborg's published diary. Walter Simon introduced me to Russell Stanton, without whom the Hanford-management section would have been incomplete. Watson Warriner provided information on the separation cells. Michael Matthias kindly gave me access to his father's files after Frank's death.

Terri Traub at DOE's Richland Reading Room and Marjorie McNinch and Barbara Hall at the Hagley Museum and Library were essential in providing access to their archives. All of them were most cooperative in responding to follow-up requests. Marjorie and Professor Seaborg pointed me to the key DuPont men mentioned above, for which special thanks. And thanks to DuPont's David Gamble for permission to use the Hagley's DuPont materials.

To the staffs of the Contra Costa County, Richmond, and Berkeley Libraries and the University of California Libraries of Engineering, Chemistry, Physics, Business and Economics, and the History offices of Boeing, the Army, the Air Force, and the Navy, thanks a lot.

A number of other Federal offices cooperated in providing supporting data. At various DOE offices, John Broderick, Mary Du Rea, Patricia Greeson, Mary Goldie, Bill Hicks, Chris Koutz, William Lynch, Robert Padgett, Larry Prete, Karen Randolph, Mike Righi, Dorothy Riehle, Yvonne Sherman, Mike Trykoski, and Tom Welch, thanks for the help. Jim Hewlett of DOE's EIA provided essential cost-index information. And thanks to Ellis Bowers and Lonnie Lewis at the Bureau of Reclamation for permission to use the Kaiser files on Grand Coulee Dam.

Tim Grogan at the Engineering News Record, Ellen Rafferty of Chemical Engineering Magazine, and Gerald Farrar of the Nelson-Farrar Index provided essential cost-index data. And thanks to Michelle Gerber for copies of her reports on the B/C reactors and the separation plant. Robert Jung of Peterson Tractor for Caterpillar information, and Tom Cooper and Fred Ackerman for permission to use the Pasco Kiwanis letter were very helpful, as was Lt. Gen R.H. Groves for permission to use an organization chart from Gen. L.H. Groves' book and Steve

Sanger for quotations from "Hanford and The Bomb." Larry Johnson provided helpful copyright advice.

Thanks also to Professor Barry Eichengreen of the UC economics Department for steering me to Madison's data on Germany. Also to Dr. Clyde Weigand, UC Physics, who provided clarification on some essential matters at wartime Los Alamos. And thanks to Don MacManman of the Tri-City Herald, Pasco, Washington for his precise definition of Horse Heaven. Also thanks to Steven Thayer for suggesting discussion of the DuPont/Met Lab relationship. My wife, Nancy, accompanied me on our trips to Richland and Wilmington and also cooperated in three years of disruption and some very large telephone bills.

The ICF-Kaiser Engineers people were their usual helpful selves. Bill Morrison reviewed the text and provided a number of valuable suggestions; Don Barrie and Alex Lindsay provided technical review. Jim McCloud and Bill Ball for their shipyard comments; Steve Armknecht for essential advice on format problems; Don Barrie, Alex Lindsay, Ralph Reynolds, Herb Thomas, and Jim Thompson for their favorable comments to the ASCE Press; Don Barrie for his guidance on current construction-management concerns; Frank Bilotti for overhead data; Dave Shrimpton for cost information; Hank Adams for cost-administration treatment; Wally Dodson and Pat Selak for questions of physics and chemical engineering, respectively; and to R.E. Bonitz, Cliff Gambs, Donna Schuske, Elaine Zacher, Mark Entes, Bea Kimsall, and Brad George for access to the Kaiser Engineers Library and Record Center for Grand Coulee data, thank you all.

Finally, with respect to publication, it wouldn't have happened without Jill Ann Martin, President of JAM Graphics, and Mary Grace Luke, Acquisitions Editor of the ASCE Press. George Stukhart and Ted Ardery of ASCE's Construction-Management Division relayed my proposed paper to the ASCE Press. Photos and permission from "The Plutonium Story" were kindly provided by Joseph Sheldrick of the Battelle Press, Stuart Loken of the Lawrence Berkeley Laboratory, and Dr. Ronald Kathren and June Markel of the Uranium Registry at Washington State University, Richland. Photos from the Hanford Science Museum collection were kindly provided by Gwen Leth and Floyd Harrow. Other essential photos were courtesy of Ray Genereaux and Russell Stanton.

Although these people contributed a very great deal of help and information, they had nothing to do with any errors of fact and interpretation committed by the author.

Chapter 14

ENDNOTES

References in this chapter are made to unidentified chapters as, "Chapters 2 and 5." References of this type are to chapters in this book.

Chapter 1. INTRODUCTION
1. Groves, p.7
2. Groves, p.127

Chapter 2. PLANT CHARACTERISTICS
1. DuPont A, p.172 says 0.025% at the top of p.172 and 0.01% at the bottom. They're the same order of magnitude, so it makes no difference for the purpose of this report.
2. Neptunium decay to plutonium: Seaborg A, pp.13-16. Reduce Iodine 131 radioactivity: Dodson
3. This map is adapted and augmented from an un-numbered page in DuPont B.
4. The pile description in this chapter is a very brief summary of the descriptions in DuPont A, pp.49-137 and Gerber A, pp.2-5 and Stanton D.
5. App.H-1, 56.-59.FM. and App.H-2, 28.WS and Gerber A, p.9 and Compton, pp.193, 194, and Dodson
6. DuPont A, p.84
7. Ibid., p.82
8. Seaborg B
9. Seaborg A, p.593 and Warriner B
10. Squires A
11. The separation-plant description in this section is a very brief summary of DuPont A, pp.172-211 and of the first 594 pages of Seaborg A. 375:1 is from DuPont A, p.199. Cell depth is from DuPont A, p.184, although 22 feet is reported in another source. The difference is inconsequential for the purpose of this report.
12. DuPont A, p.52
13. App.H-1, 44.FM
14. Jones, p.401
15. Hageman, p.69
16. Jones, p.110 and Dupont A, p.49 and scaled takeoff from Map No. 4, Jones

17. DuPont B, v.4, from which the author summed the total from building descriptions
18. Same as note 17
19. DuPont A, p.49 and App.G
20. DuPont B, v.4, from author's total from bldg. descriptions
21. Hageman's p.69 total minus Note 20 quantity
22. Carpenter A, p.4
23. DuPont B, v.4, from which the author summed from bldg. descriptions
24. Hageman, p.70
25. DuPont B. v.4
26. Matthias D, p.118
27. Ibid.
28. Ibid.
29. Hageman, p.69
30. Carpenter A, p.4
31. Matthias B, p.51
32. App.H-1, 54.FM and Carpenter A, p.4
    In Matthias B, p.155, the graph of Construction Force On Roll shows 44,000 peak, but the majority of sources use 45,000.
33. Hageman, p.70

Chapter 3. INITIAL UNCERTAINTIES AND DIFFICULTIES

1. "Never in history": Groves, p.72
   "DuPont's Harrington": Ibid., p.47
   "Even after DuPont": Compton, p.133
   "Seaborg remarked": Seaborg A, p.82
   "Both the Corps and James": DuPont A, p.14
   "An indication of our": Seaborg A, p.196
2. "When the corps was directed": DuPont A, p.10
   "Groves told the DuPont": Ibid., p.15
   "We had no sure knowledge": Compton pp.27-29 and App. H2, 14-16.WS
   "We could not afford to wait": Groves, p.72
   "Nuclear physics and": DuPont A, p.4
   "It was considered possible": Ibid., p.4
   "DuPont had required a total": Hounshell & Smith, pp.233-273
   "The engineering difficulties": Groves, p.20
   "There was no advance": Author's conclusions from the literature and interviews
3. "Separation-process design": Seaborg A, pp.365 and 614, and DuPont B Building Facilities progress charts for 200 W
   "Final design of the separation": Chapters 5 and 6
   "All of the remote-handling": Chapter 5
   "Remote replacement": Chapter 5

"A large number of key": DuPont A, pp.319-413
4. Compton, pp.90-99

Chapter 4. THE MANHATTAN EFFORT
1. NDRC established: Groves, pp. 7, 8 and Nichols, p.33
   OSRD established: Nichols, p.33
   Bush recommends Corps: Jones, p.37
2. Groves, p.9
3. App.H-1, 127.FM
4. Groves, pp.10, 417 and Smyth p.83
5. Groves, p.2
6. App.H-1, 127.FM
7. Matthias C, p.8
8. Jones, p.77
9. Matthias interview, plus his personal files, plus interviews with Simon, Genereaux, and Stanton. Also Groves, p.2 and Seaborg A
10. Groves, p.22
11. Ibid., p.28
12. "go where the work was," Nichols, p.39
13. Nichols, p.106
14. Ibid., p.38
15. Groves, p.28
16. Matthias C, p.13
17. App.H-1, 130.FM
18. Matthias E
19. App.H-1, 96, 98, 100.FM
20. Ibid., 117.FM, 126.FM
21. Ibid., 82, 94.FM
22. App.H-5, 60.RS
23. Matthias C, p.13
24. App.H-1, 83, 84, 92, 95.FM
25. Nichols, p.60
26. Nichols, p.60, and Matthias-Thayer conversation, 1993

Chapter 5. THE DUPONT EFFORT
1. Seaborg A, pp. 173, 237 and DeRight, p.1
2. Seaborg A, p.237
3. Groves, pp.42,43
4. DuPont A, pp.10, 11 and DeRight, pp.1, 2 and Groves, p.51
5. Hounshell & Smith, p.338
6. Chandler & Salsbury Pierre's degree, p.17; Executive Committee, p.129
7. Hounshell & Smith, p.286
8. Ibid., p.120

9. Ibid., Nylon development and Carothers' role, pp.233-274
10. Ibid., pp.275-285
11. Genereaux A, p.6
12. Hounshell & Smith, p.266
13. Ibid., p.359
14. Seaborg A, p.217
15. Hagley Museum and Library, Acc. 1957, Box1, File 2
16. Perry, pp.v-ix, and flyleaf
17. Genereaux A, pp.12, 14, 15
18. Hounshell & Smith, pp.184-189 and DuPont A, "Appendix" following main page 463
19. Lewis History, 1948
20. Read, pp.1, 2
21. Hounshell & Smith, p.333
22. App.H-1, 25.FM and Groves, pp.42, 43
23. App.H-2,1-3.WS, 18.WS, 20.WS; App.H-3, 1.RG, 8.RG, Genereaux P; App.H-4, 11.JT, 12.JT, 16.JT; App.H-5, 14.RS, 29.RS, 60.RS, Stanton B; October, 1993 phone conversation with Simon
24. Hounshell & Smith, p.607
25. Ibid., p.122 shows 7 auxiliary departments in 1931 and 12 in 1961. I have found no source for the 1941 auxiliaries.
26. DuPont A, "Appendix" following p.463, pp.v,vi. Fig. 5.2 is my simplification of the Appendix data, as supplemented by App.4, 10.JT and Squires B.
27. App.H-1, 28.FM; App.H-5, 13.RS
28. DuPont D. Figure 5.3 is my simplification of this chart. DuPont A, p.275, states the existence of a "Central Procurement Section in the Design Division." The above-referenced organization chart does not show this section. I do not believe it existed, because it would have been unnecessary. The functions that the p.275 author had in mind – specifications writing and bid review – are part of the normal Design Division function.
29. Dutton, pp. 226-245
30. Carpenter A, p.2
31. Seaborg A, p.210
32. Ibid., pp. 116, 156, 178, 241, 265, 335, 341, 576, 593
33. Ibid., p.365
34. Barrie & Paulson, pp.43, 64
35. Genereaux P
36. App.H-2, 14.WS
37. App.H-4, 9.JT
38. Genereaux N
39. Genereaux K
40. App.H-2, 11.WS
41. DuPont A, p.356
42. Seaborg A, p.471 and DuPont B, building and facilities progress charts

43. DuPont A, p.204
44. App.H-2, 20.WS, plus a lunchtime conversation with Walter Simon on Sept. 23, 1993
45. "total and constant attention to detail," Sanger, p.42 and Genereaux F

    "did not spare good design...", App.H-3, 20.RG

    "...leaving it to pipefitters'..." App.H-3, 9.RG

    "...operations people worked w/engineering..." App.H-2, 3, 8.WS

    "Elaboration was avoided..." Genereaux P, and Sanger, p.42

    "...as the ultra-high-pressure..." App.H-3, 8.RG and App.H2, 20.WS
46. Genereaux F and App.H-3, 13.RG and Graves, p.30 and Sanger p.148
47. Squires A and App.H-3, 27, 28 RG and Gerber B, p.4
48. App.H-2, 5.WS and App. H-3, 7, 21, 24, 34.RG
49. App.H-3, 11, 12.RG, and conversation with Matthias prior to interview.
50. DuPont A, pp. 36, 37
51. App.H-3, 29-32.RG and Genereaux N
52. This account of separation-cell design was from App.H-3, 7.RG and Genereaux E, F, and N and Squires B.
53. App.H-3, 20, 26.RG
54. App.H-3, 7, 20, and 26.RG and Genereaux K
55. Equipment selection and special sensors: DuPont A, 241
56. Assembly & disassembly: App.H-3, 7.RG and Genereaux I
57. Remote installation and Construction resistance: Genereaux K and App.H-3, 18, 20.RG
58. App.H-3, 18, 20.RG
59. Construction joints: Warriner
60. Genereaux K
61. Rhodes, p.560 and Graves, p.32 and Compton, pp.193, 194, and Seaborg C
62. App.H-5, 50.RS and Nichols, p.215 and Hewlett & Anderson, p.334 and Compton, pp. 267-271
63. Sanger, p.60 and App.H-1, 102-104.FM and App.H-5, 55.RS
64. App.H-2, 17.WS and App.H-3, 22, RG and App.H-4, 18.JT

    The count of over 40 items and the examples cited were compiled from DuPont A, pp.69-391. Non-Hanford work, DuPont C, p.15
65. 47,304 P.O.s; $2.1 billion; 47 states: DuPont B, pp.223-225 and App.G for escalation

    74 subcontracts; $1.1 billion: DuPont A, pp.292, 293 and App.G for escalation.

    DuPont placed 66.8% of total purchase orders in the 11 western states. 11.2% of Wilmington orders were placed in the 11 western states. These percentages were calculated from the tabulation in DuPont B, pp.224, 225.
66. Space in vendors' schedules: DuPont A, pp.307, 308, 349. Waive competitive bidding, splitting orders, orders concurrent: DuPont A, pp.278-296.

    Buying machine tools: DuPont A, pp.324, 326

    Procure used equipment: DuPont A, p.314

    Buy from warehouse stock: DuPont A, pp. 308-347

67. Slug and block groups: DuPont A, pp. 319-413
68. Block configuration and tolerance: DuPont A, pp.321, 322
69. These blocks comprised...to end of par.: Compiled from data in Fig. 5.7

    14 cities is an approximate number compiled from Thomas, which listed a number of factory cities for several of the vendors in 1942. In those cases, I would pick one. Chicago and Pittsburg were common duplicates. Thus the number of cities is smaller than the number of vendors. Figures 5.6, 5.7, 5.8, and 5.9 are Hagley Acc. 1957, Box 58/26/7-11 and Acc. 1463, v.1.
70. Block group organization: DuPont A, pp.320, 321
71. 70 fabrication and admin. problems: Compiled from DuPont A, pp.323-339
72. Procurement dates: DuPont A, p.323 and Fig. 5.6
73. 82 vendors and 129 problems: Compiled from DuPont A, pp.319-413
74. Genereaux D
75. Stainless order: DuPont A, pp.309-313
76. 5800 people were counted from the organization charts of DuPont B (See Endnote 77) I queried Mr. Stanton about his total force (App.H.5, 42.RS) and determined that he had 80 more people than the 12 shown in his group in Fig. 5.10. I therefore augmented the forces under the other Division Engineers under Grogan in proportion to the Stanton augmentation.
77. DuPont B, v.1. These six charts were on un-numbered pages in the copies at the Hagley and on the corresponding blank pages at DOE in Richland.
78. Buses: Stanton B
79. Location of scheduling effort on charts: Stanton A
80. This account of the origin of CPM and its use at Hanford is to be found in none of the literature. The sources are in the following: App.H-1, 29.FM; App.H-3, 2-6.RG; App.H-5, 14, 16, 17.RS and Stanton E and I.
81. This engineer-supervision method does not appear in the literature. The source is in the following interviews: App.H-1, 30-37.FM; App.H-5, 1, 2, 4, 9, 10, 13, 42, 43, 53.RS.
82. This section on DuPont's QA is taken from: App.H-1, 101.FM; App.H-2, 31, 32.WS and 37.WS; App.H-4, 14, 15.JT; App.H-5, 34-38.RS; Stanton A-D, L and M.
83. App.H-1, 101.FM; App.H-2, 21-23.WS, and App.H-5, 56-59.RS
84. Seaborg's remark: Seaborg B; procedure signoff: Seaborg A, p.579

Chapter 6. THE METALLURGICAL LABORATORY EFFORT

1. Compton, p.86, and Groves, p.9

    Met Lab control was transferred to the Manhattan Project from the S-1 Committee on Feb. 17, 1943 (Seaborg A, p.249)
2. Tasks 1, 2, 4, 5, 6, 11: Seaborg A, pp. 160, 237, 241, 244, 251, 255, et.al.

    Tasks 5, 7.8, 10: Seaborg A, p.116

    Task 9: Seaborg A, p.153

    Task 11: Seaborg A, pp.193, 195, 196, and DuPont A, p.82

    Task 3: DuPont A, pp.61-392

    > Met Lab as central coordinator of slug program: DuPont A, p.358

      Task 12: App.H-1, 104.FM
      Task 13: App.H-5, 34.RS
3. Bomb designed detached: Seaborg A, p.249
4. Simplified chart taken from Seaborg A, Tbl.1, p.237
5. Seaborg A, pp.90, 109
6. Ibid., p.614
7. Ibid., pp.116, 156, 178, 241, 265, 335, 341, 576, 593
8. Squires C
9. DuPont B Design Status charts & Progress-of-Construction charts
10. First shipment: Seaborg A, p.365; 30 unsolved problems,
      Seaborg A, p.398
11. March 6-12 principal problems: Seaborg A, pp.421, 422
12. Criticality in 231 Bldg.: DuPont A, p.203; June 20 extraction changes: Seaborg A, p.471; Mechanical isolation changes in 231: DuPont A, p.203; change to air jets, DuPont A, p.204
13. Carpenter A, pp.2, 3
14. App.H-2, 3.WS
15. Drawing percentages: DuPont B, Design Status charts; cell building superstructure: DuPont B, Building Facilities completion record, 200 West
16. Seaborg A, p.209
17. App.H-2, 28.WS. For the reactor tube controversy, see Chapters 2 & 5

Chapter 7. THE COSTS OF THE HANFORD WORKS
1. DuPont A, pp.435, 436
2. Hewlett, James, A and B
3. Carpenter B, p.2 and App.H-1, 67, 68.FM
    The Kiwanis Club letter is courtesy Mr. W. Thomas Cooper, President of the Pasco Kiwanis Club, Mr. Fred Ackerman of the same organization, and of the Hagley Museum & Library.
4. Hewlett and Anderson, p.2
5. Department of Commerce, v.1, pp.221-228
6. $51.9 billion of public and private construction and $8.5 billion of military facilities: Department of Commerce, v.2, pp.618, 619
    8,948 naval vessels: Department of the Navy, 1945
    88,410 tanks: Department of the Army, 1950
    324,750 airplanes: Department of the Army, U.S. Army Special Production Studies
7. One quarter of the U.S.: Madison, p.160, 161
    Cyclotron 12 years later: Speer, p.228 (German cyclotron, 1944) and Ado and Shoemaker, Encyclopedia Britannica (Lawrence's cyclotron, 1932)
8. DuPont A, pp.309-350
9. Ibid., p.330

Chapter 8. LABOR, CONDITIONS
1. Matthias B, p.51
2. App.H-1, 79.FM
3. App.H-5, 4.RS
4. Matthias B, p.52
5. Jurisdictional agreement; interchange of crafts; doubt that work could have been completed on schedule: Matthias B, pp10 and 50
6. App.H-1, 80.FM
7. Ibid., 81.FM
8. Ibid., 77-79.FM
9. Ibid., 118.FM
10. App.H-5, 40.RS
11. Ibid., 52.RS
12. Matthias B, p. 29
13. App.H-1, 119.FM
14. App.H-5, 30, 31.RS
15. Parker, p.15
16. App.H-5, 30.RS
17. App.H-1, 70-74.FM
18. App.H-2, 33.WS
19. Matthias B, p.23
20. Ibid., pp.5, 23
21. Ibid., p.18
22. "Major Injury Chart," (2-18-45, for '43, '44, &'45) Hagley file 1-c
23. Matthias B, p.55

Chapter 9. TECHNOLOGIES OF THE FORTIES.
1. Jung
2. Warburton
3. App.H-1, 3.FM, 18.FM. App.H-2, 25.WS
4. App.H-3, 17.RG
5. Seaborg A, pp.94-155

Chapter 10. INTANGIBLES
1. App.H-2, 13, 20.WS, and App.H-3, 14.RG, and App.H-4, 5, 6.JT, and App.H-5, 48.RS
2. App.H-2, 35.WS, and App.H-1, 131.FM
3. App.H-2, 14, 16.WS, and Seaborg A, p.158
4. Seaborg A, p.217
5. App.H-5, 60.RS
6. Compton, pp.94, 95
7. Groves, pp.47, 48
8. App.H-1, 24-26.FM

9. Squires A
10. Compton, pp.8, 73, 78, 85
11. Matthias B, pp.16, 47
12. App. H-2, 16.WS
13. Parker, pp.3, 9
14. Welch

Appendix A

LIST OF FIGURES & TABLES

Figures — Page No.

2.1	The Hanford Engineer Works (Map)	6
2.2	Block Spacing	7
2.3	Diagrammatic Pile Cross Section	10
4.1	Manhattan Project Simplified Organization Chart	22
4.2	Manhattan Project – Hanford Related Elements	23
4.3	Manhattan/Hanford – Simplified Organization Chart	28
5.1	Crawford Greenwalt's Notes of the First Sustained Nuclear Chain Reaction In History	33
5.2	Explosives Department TNX – Simplified Organization Chart	36
5.3	Engineering Department – Simplified Organization Chart	37
5.4	Parallel Development Events vs. Drawings, Manpower, and Construction Progress	44
5.5	Parallel And Rational Schedules	46
5.6	Shield Block Completion, Acceptance, And Shipping Chart	57
5.7	Biological Shield Block Weekly Flow Chart	59
5.8	Reactor Nozzle Flow Chart	60
5.9	Control Card For Sleeves	61
5.10	The Hanford Engineer Works – Simplified Field Management Organization	64
6.1	The Metallurgical Laboratory – Simplified Organization Chart	77
7.1	Pasco Kiwanis Club Letter to Walter S. Carpenter	85
8.1	Hanford Engineer Works – Construction Force	94

Tables

2.1	Separation-Process Sequence	12
2.2	Hanford Works Quantities	17
5.1	Prominent Questions Paralleling Design and Construction	43
5.2	Hanford Design – Development Programs Charged to The Wilmington Engineering Office	49
5.3	Special Fabrication Problems – Laminated Blocks	62
5.4	Hanford Construction Organization Branch Functions & Management Relationships	65
5.5	Examples of Construction Procedures	71
7.1	The Recorded Costs of Hanford	82
7.2	Cost Indices and Multipliers	83
7.3	Manhattan And HEW Costs vs. U.S. Gross National Product	86
7.4	Comparison of Adjusted M&E Costs	87

Appendix B

ABBREVIATIONS & ACRONYMS

This list is limited to those items not necessarily familiar to all engineers.

ANSI/ASME NQA-1	"Quality Assurance Program Requirements For Nuclear Facilities," a publication of ANSI and the ASME; the base, governing document for nuclear quality assurance
CMX	The water corrosion and treatment experiment at Hanford (Ch. 2)
GAO	General Accounting Office
HEW	Hanford Engineer Works
k	The multiplication factor in nuclear physics. In fission, the ratio of the production of neutrons to the loss of neutrons. Must be greater than one for a chain reaction to occur.
MED	Manhattan Engineer District[a]
Met Lab	The Metallurgical Laboratory at the University of Chicago[a]
NDRC	National Defense Research Council[a]
NQA-1	See ANSI/ASME NAQ-1.
OIC	Officer In Charge
OSRD	Office of Scientific Research And Development[a]
S-1	The OSRD branch for the atomic effort[a]
SMX	The non-nuclear experimental pile at Clinton (Ch. 2)
TNX	Explosives-TNX; a DuPont department (Ch. 5)
UC	The University of California at Berkeley
100, 200, 300 Areas	The reactor, separation, and experimental/manufacturing areas at Hanford, respectively
$^{239}_{94}$	The notation for $^{\text{atomic weight}}_{\text{atomic number}}$ of a chemical element, in this case, plutonium 239

---

[a] See Ch.1, Introduction

Appendix C

GLOSSARY

This list is restricted to those items not necessarily familiar to all engineers.

Can	The aluminum jacket enclosing the uranium slug to protect the uranium from water and to confine the fission products
Canyon	The 221 separation-cell building
Crafts/Craft	A noun or adjective, used in construction vocabulary with the following meanings: a particular construction trade, as carpenter, machinist, pipefitter, etc.; workers, or groups of workers in a trade, or all of the trades; the aggregate of all trades of all types
Decontamination (1944 usage at Hanford)	The chemical process of removing undesirable fission products from the separation-process stream
Dummy slug	Spacer used in the fuel tube to place the fuel slugs in the locations required by reactor physics
Fission products	Any nuclides, either radioactive or stable that arise from fission including both the primary fission fragments and their radioactive decay products.
Horse Heaven	The area of southeastern Washington lying principally on the north slope of the Columbia River between the Wallula Gap and Bickleton, but extending to the Yakima River. So named because the late 19th-century stands of high bunch grass made a heaven for the wild mustangs. Now dry-land farming and horse ranches.
Lattice	Spacing of fuel elements in the pile
Lewis Committee	A group appointed by Groves to: compare candidate processes for production of bomb material; determine which had the best chance for success and which should be built. The members: Roger Williams, Crawford Greenwalt, and T.C. Gary of DuPont; and from Standard Oil, Eger Murphree. Headed by W.K. Lewis of MIT.
Microgram	Millionth of a gram
Milligram	Thousandth of a gram
Moderator	A material used in a reactor to slow the neutrons to prevent their escape before they can chain react
Pile	The stack of graphite blocks penetrated by fuel pipes containing uranium slugs to cause the chain reaction
Poisoning	Excessive absorption of neutrons, thus stopping the chain reaction. The fission product Xenon poisoned the Hanford B Reactor upon initial startup. Intended poisoning is by control rods.
Slug	The uranium fuel element in the fuel tubes in the pile. 8 inches long by about 1.5 inches diameter. Enclosed in the can.
Spool	A piping sub-assembly consisting of pipe and fittings, preassembled for later inclusion in a completed piping system

Appendix D

PERSONNEL LIST

This is a list of principal figures and those mentioned repeatedly in the text. Figures appearing in organization charts are not listed here, except as they fit the categories of the first sentence.

Ackart, E.G. – Chief Engineer of DuPont's Engineering Department

Bush, Vannevar – Head of the NDRC; former Vice President of MIT; developed first analog computer; founder of Raytheon; member, National Academy of Sciences; adviser to President Roosevelt.

Carpenter, Walter – President of DuPont. Prominent in management of DuPont's mammoth WW I powder-plant expansion.

Church, Gilbert P. – DuPont's Manager of Hanford Construction

Compton, Arthur H. – Head of the Met Lab; Chairman of the Physics Department of the Univ. of Chicago; Chairman, National Academy of Sciences committee on military uses of uranium.

Conant, James Bryant – Head of the S-1 Committee; President of Harvard; Former Chairman of Harvard's Department of Chemistry.

Cooper, Charles – Director of Technical Division of the Met Lab; managed the chemical-engineering for Hanford's separation process; former Director of MIT's School of Chemical Engineering; chemical engineer for DuPont, who loaned him to the Met Lab for the plutonium project.

Fermi, Enrico – Director, Nuclear Physics Division of the Met Lab; managed design and construction of the pile which generated the first sustained nuclear chain reaction in history; managed conceptual design for Hanford's reactors.

Genereaux, Raymond P. – DuPont's Assistant Design Project Manager for separation plant design; former Head, Chemical-Engineering Branch of DuPont's Scientific Research Group; Only five-time contributor to the Chemical Engineer's Handbook.

Graves, George D. – Assistant Manager of the Technical Division of Explosives TNX; insisted on an additional 504 fuel tubes in the Hanford reactor design; prominent in DuPont's nylon process design.

Greenwalt, Crawford – Manager of the Technical Division of Explosives TNX; former Research Director of DuPont's Grasselli Division; member, Board of Directors; subsequently President of DuPont.

Groves, Major Gen. Leslie R. – Head of the Manhattan Project; graduate of West Point; former Deputy Chief of Army construction; former manager of construction for the Pentagon.

Lawrence, Ernest O. – Director of the UC Radiation Laboratory; directed development of the magnetic separation process for U 235; inventor of the cyclotron.

Marshall, General George C. – Chief of Staff of U.S. Army; Groves' immediate superior for military application of the bombs.

Matthias, Col. Franklin T. – Officer In Charge at Hanford; former Deputy Manager of construction for the Pentagon; subsequently managed major construction projects as civilian; subsequently Vice President, Heavy Construction for Kaiser Engineers.

Nichols, Col. Kenneth D. – District Engineer, Manhattan District; managed Oak Ridge program directly, and administrative director for the entire District; held Ph.D. in hydraulic engineering; subsequently General Manager, U.S. AEC.

Reybold, Lt. Gen. Eugene – Chief of Engineers under Gen. Marshall. Provided assistance to the Hanford project in furnishing advice, equipment, and personnel, but voluntarily withdrew from direct supervision.

Seaborg, Glenn T. – Section Chief, Final Products (Plutonium) in the Chemistry Div. of the Met Lab; co-discoverer of plutonium as faculty member of UC; directed development of plutonium chemistry at the Met Lab as basis for chemical engineering of Hanford's separation process; subsequently head of the U.S. AEC and Chancellor of the University of California.

Simon, Walter O. – Hanford's plant operations Manager; subsequently DuPont's Buffalo Plant Manager.

Squires, Lombard – Head of separation-plant process coordination in the Technical Division of TNX; former instructor in chemical engineering at MIT; subsequently member of the Advisory Committee on Reactor Safeguards of the AEC.

Stanton, Russell C. – Division Engineer in charge of reactor construction at Hanford; former construction engineer on DuPont projects. Subsequently Assistant Field Project Manager at Savannah River.

Stimson, Henry L. – Secretary of War; Groves' immediate superior.

Stine, Charles M.A. – Director of DuPont's Chemical Department and Vice President of DuPont. Member of the Managing Committee for McGraw-Hill's Chemical Engineering Series.

Tepe, John B. – Research chemist loaned by Dupont to the Met Lab for development of the chemical separation process.

Wigner, Eugene P. – Section Chief, Theoretical Physics in the Nuclear Physics Division of the Met Lab; former Thomas D. Jones Professor in Mathematical Physics at Princeton.

Williams, Roger Sr. – Assistant General Manager in charge of Explosives TNX; former chemical Director of DuPont's Ammonia Department; managed high-pressure systems technology at DuPont's Belle, West Virginia plant.

Appendix E

REFERENCES

The materials found at the Hagley Museum And Library are from their Manuscripts And Archives Department and are courtesy of the Hagley Museum and Library.

The DuPont Design and Construction Histories are available at both the Hagley and at the DOE Public Reading Room in Richland, Washington. The copies at the latter often have blank pages in place of the charts.

Matthias Items B,C, and E were found at his home in Danville, California, where I photocopied them courtesy of his son Michael. Subsequently, Michael donated Col. Matthias' entire Hanford collection to the Hagley Museum and Library, where it has been given Accession No. 2086.

Col. Matthias' diary is courtesy of the DOE Public Reading Room.

The locations of the unpublished sources listed herein are:

DOE Public Reading Room	WSU-Tri Cities, Room 130 100 Sprout Road, Mail Stop H2-53 P.O. Box 999 Richland, WA 99352
Hagley Museum And Library	P.O. Box 3630 Wilmington, DE 19807
ICF Kaiser Engineers	1800 Harrison St. Oakland, CA 94612

Appendix E

REFERENCES

American Society of Civil Engineers. *Manual No. 45. A Guide For the Engagement of Services.* New York: 1968

———. *Manual No. 45. A Guide For the Engagement of Services.* New York: 1988

Barrie, Donald S. and Boyd C. Paulson. *Professional Construction Management.* New York: McGraw Hill, 1992

Carpenter, Walter S. Jr. Letter to DuPont employees, Aug. 24, 1945. Hagley Museum and Library, Carpenter papers, II/2/Box 830
CARPENTER A

———. Comments At Presentation of Army-Navy "E" to Hanford Plant, Oct. 9, 1945. Hagley Museum and Library, 11/2/Box 830
CARPENTER B

Chandler, Alfred D. and Stephen Salsbury. *Pierre S. DuPont and The Making of the American Corporation.* New York: Harper & Row, 1971

Compton, Dr. Arthur H. *Atomic Quest.* New York: Oxford University Press, 1956

Consolidated Builders Inc. (CBI). Records of Grand Coulee Dam. Files of ICF Kaiser Engineers, Oakland, Calif.

Department of the Army, 1950. Army Comptroller's Report. Filed at the Center For Military History, Washington, D.C.

———. U.S. Army Special Production Studies, by Irving Holley. Filed at the Air Force History Center, Washington, D.C.

Department of Commerce, 1975. *Historical Statistics of The United States - Colonial Times to 1970.* Washington, D.C.: GPO

Department Of Energy, 1985. *Department of Energy Acquisition Regulation (DEAR), DOE/MA-0189.* Washington, D.C.: NTIS

Department of The Navy, 1945. Secretary of The Navy's Report. Filed at Navy History Center, Washington, D.C.

DeRight, Robert. Time Table. Hagley Museum and Library. (This is a chronological set of notes concerning DuPont's plutonium-related activities, kept from July 14, 1942 to August 26, 1943, and is dated Sept. 8, 1943.) Hagley Acc. 1957/I/A, Box 1,/2

Dodson, Wallace J. Telephone call record, Aug. 16, 1993 providing details of Uranium transformation to Plutonium, and the decay of Iodine 131. Dodson is the retired head physicist of the Nuclear Division of Kaiser Engineers.

DuPont DeNemours, E.I. And Company (Inc.). Design and Procurement History of Hanford Engineer Works. Hagley Acc. 1957.
DU PONT A

———. Construction, Hanford Engineer Works. (In four volumes)
DU PONT B Hagley Acc. 1463,1957.

_____. Engineering Department Heritage, Part 5, WW II Military Construction. In-house brochure, undated. Hagley Museum and Library, Acc. 1957
DU PONT C.

_____. Organization Chart, Engineering Department at Wilmington, June 10, 1944. Hagley Museum and Library, Acc.1463,V.1
DU PONT D

_____. Production tracking charts Hagley Acc. 1957, Box 58, File 26, and Acc. 1463, V. 1
DUPONT E

Dutton, William S. DuPont - One Hundred and Forty Years. New York: Charles Scribner's Sons, 1951.

Genereaux, Raymond P. Interview by John K. Smith, Oct. 28, 1982. Hagley Museum and Library, Acc. 1878
GENEREAUX A

_____. Telephone record, April 25, 1994 re: separation group's not needing special procurement group. Author's file.
GENEREAUX D (In the alphabetical Genereaux series, only those items used in the book are listed in the References.)

_____. Telephone record, June 3, 1994 re: cell piping design. Author's file.
GENEREAUX E

_____. Letter to Author, August 4, 1994, consisting of markup of July 27 draft of Engineering section of Ch. 5. Author's file.
GENEREAUX F

_____. Letter to Author, Dec. 4, 1994 providing his comments on the book's complete typescript. Author's file.
GENEREAUX I

_____. Telephone record, June 13 and 27, 1995 re: remote assembly of and 3 crucial keys to separation-plant design and construction. Author's file.
GENEREAUX K

_____. Telephone record, July 17, 1995 re: Clinton semi-works, exempt overtime, design secrecy. Author's file
GENEREAUX N

_____. Engineers, Engineering, And The Automatic Factory. Typescript of a paper presented at the ASME Annual Meeting, New York City, Dec. 1, 1953. Photocopy in Author's file.
GENEREAUX P

_____. Telephone record, August 14, 1995 re: hot-cell assembly and components. Author's file.
GENEREAUX Q

Gerber, Michelle. Summary of 100 B/C Reactor Operations and Resultant Wastes, Hanford Site. Richland WA: Westinghouse Hanford Co., 1993. Westinghouse Hanford Co. Central files. WHC-SD-EN-RPT-004.
GERBER A

_____. A Brief History of The T Plant Facility At The Hanford Site. Richland, WA: Westinghouse Hanford Co., 1994. Westinghouse Hanford Co. Central files. WHC-MR-0452.
GERBER B

Graves, George D. Interview by John K. Smith, May 24, 1983, Hagley Museum and Library, Acc. 1878

Groves, Lt. Gen. Leslie R. Now It Can Be Told. New York: Harper & Brothers, 1962.

Hageman, Roy C. Threshold of An Era. Hanford, WA, 1945, Hagley Museum and Library. Acc. 1957, Box 1

Hewlett, James G, Robin A. Cantor, and Colleen G. Rizzy. An Analysis of Nuclear Power Plant Construction Costs. Washington, D.C.: Energy Information Administration, Office of Coal, Nuclear, Electric, and Alternate Fuels, U.S. Department of Energy.
HEWLETT A

_____. Telephone record, July 10, 1995 re: nuclear cost escalation data. Author's file.
HEWLETT B

Hewlett, Richard G. and Oscar E. Anderson Jr. The New World. WASH 1214, v.1, 1972 USAEC, Washington, D.C.: U.S. Atomic Energy Commission, 1972

Hounshell, David A. and John Kenly Smith Jr. Science And Corporate Strategy: DuPont R&D, 1902-1980. New York: Cambridge University Press, 1988

Jones, Vincent C. Manhattan: The Army And The Bomb. Washington, D.C.: Center for Military History, United States Army, 1985.

Jung, Robert. Telephone record, Aug. 2, 1995 re: tractor specifications (Peterson Equipment Co.)

Lewis History Publishing Co. Who's Who In Engineering. New York, 1948

Madison, Angus. Phases of Capital Development. New York: Oxford University Press, 1982

Matthias, Col. Franklin T. Diary for the years 1942-1945. The diary begins with 37 handwritten pages through Jan. 29, 1943 and then is typescript to the end. This is a copy of the original. DOE Reading Room.
MATTHIAS A

_____. Procurement Of Labor. A set of 87 unnumbered but sequential pages of typescript and photostatted documents describing HEW labor conditions, hand dated October, 1945, and initialed "FM" in lower right corner of page 1. I photocopied my set from the original, at that time in Col. Matthias' home in Danville, Calif., and hand numbered the pages. Author's file and Hagley Museum and Library.
MATTHIAS B

_____. Concept, Design, And Construction of the First Large Nuclear Reactor, Hanford, 1943-1945. A typescript draft of a paper presented at the spring convention of the American Society of Civil Engineers, Pittsburg, Pa, 1978. Author's file (photocopy). Hagley Museum and Library (photocopy).
MATTHIAS C

_____. "Building the Hanford Plutonium Plant." Engineering News Record, Dec. 13, 1945, p.118.
MATTHIAS D

_____. Organization chart for the Manhattan District at Hanford Engineer Works; 9pp. I supplemented this chart by an expanded chart of the Hanford-Manhattan Construction Group under Lt. Col. Rogers found at the Hagley Museum and Library.
MATTHIAS E

McCullough, David. The Path Between The Seas. New York: Simon and Schuster, 1977.
MC CULLOUGH A

_____. Truman. New York: Simon and Schuster, 1992.
MC CULLOUGH B

Nelson, Robert W. Telephone record, Mar. 5, 1994 re: UC purification of uranium chloride from Met Lab. Author's file.

Nichols, Col. Kenneth D. The Road to Trinity. New York: Morrow, 1987.

Parker, James. Untitled 15pp. typescript memoir of Mr. Parker's life at Hanford and his work in electrical construction for three contractors in the Tri-Cities area, including Newberry, Chandler, and Lord, the HEW electrical contractor. Author's file.

Perry, John H., Editor. Chemical Engineers' Handbook, Second edition. New York: McGraw Hill Book Company, Inc. 1941.

Read, Granville M. Letter to E.G. Ackart, Nov. 2, 1942. Resume of War Construction Activities. Hagley Museum and Library. Acc. 1463

Rhodes, Richard. The Making of The Atomic Bomb. New York: Simon and Schuster, 1988.

Sanger, S. L. Hanford and The Bomb. Seattle: Living History Press, 1989.

Seaborg, Professor Glenn T. The Plutonium Story. Edited by Ronald Kathren, Jerry B. Gough, and Gary T. Benefiel. Columbus: Battelle Press, 1994.
SEABORG A

_____. Telephone record, January 3, 1994. Pu discovery and confirmation; his opinion of DuPont's Hanford work.
SEABORG B

_____. Telephone record, August 2, 1994. Wigner's memo to Compton; Graves and the 2004 fuel tubes. Author's file.
SEABORG C

Smyth, Henry DeWolf. Atomic Energy For Military Purposes. Princeton: Princeton University Press, 1945.

Speer, Albert. Inside the Third Reich. New York: Collier Books Div. of Mac Millan, 1970.

Squires, Lombard. Telephone record, April 4, 1994. Separation process design. Author's file.
SQUIRES A

_____. Telephone Record, June 17, 1994. Separation process design. Author's file.
SQUIRES B

_____. Telephone record, Oct. 22, 1994. TNX/MET Lab relationship. Author's file.
SQUIRES C

Stanton, Russell C. Telephone record, Mar 21, 1994 re: DuPont Hanford organization; startup; surge chambers. Author's file.
STANTON A

_____. Telephone record, Sept. 1, 1994 re: OIC organization; reactor weld inspection. Author's file.
STANTON B

_____. Letter to author, Sept. 29, 1994, transmitting his markup of August 19 draft of the Hanford Field Management section of Ch. 5. Author's file.
STANTON C

_____. Telephone record, Oct. 13, 1994 re: construction details, costs, and QA. Author's file.
STANTON D

_____. Telephone record, Feb. 1, 1995 re: history of the Critical Path Method at DuPont. Author's file.
STANTON E

_____. Telephone record, July 26, 1995 re: samples of procedures titles, his job chronology, SMX chronology, and DuPont's 1948 return to Hanford. Author's file.
STANTON I

_____. Stanton letter of Aug. 30, 1995 re: sample list of reactor QA procedures. Author's file.
STANTON L

_____. Telephone record, August 30, 1995 re: pile erection procedure. Author's file.
STANTON M

Thomas Publishing Co. *Thomas Register*, New York: 1941

Warburton, R.L. p.76 of a DuPont report of inventory accounting at Hanford. Hagley Museum and Library. Acc. 1957, Box 50

Warriner, Watson. Telephone record, Aug. 25, 1995 re: separation-plant design. Warriner was a separation-plant design engineer under Genereaux, and then went to Hanford to monitor separation-plant construction.
WARRINER A

_____. Letter to author, Dec. 26, 1995 correcting building numbers in Table 2.1.
WARRINER B

Welch, William. Telephone record, Sept. 16, 1993. Mr. Welch's experience as an expediter at HEW from Spring to Fall, 1943. Author's file.

Appendix F

CHRONOLOGY OF THE HANFORD PROJECT

The two purposes of this chronology are:

As an overview for those seeking a broad view of the progress of the Hanford program. The overview events are indicated by dates in bold-face type.

As an illustration of the concept of parallel development by providing the dates of detailed events in physics, chemistry, conceptual and final engineering, semiworks, procurement, construction, and operations.

Events at Hanford are indicated by an asterisk to the right of the date. Procurement-related events are indicated by a "P" to the right of the date. As stated in the Introduction for this report as a whole, this is not intended as a chronology of either nuclear physics, radio chemistry, the total plutonium project, or the Manhattan project.

June,	1940	National Defense Research Committee established; Vannevar Bush in charge. Became subsidiary to OSRD upon its establishment. (Groves, p.7, 8; Nichols, p.33)
Dec.	**14, 1940**	Plutonium 238 created by Seaborg, Kennedy, and Wahl in UC 60-inch cyclotron. Confirmed chemically on night of Feb. 23, 24, 1941. (Seaborg phone call, Jan. 3, 1994 and Seaborg A, pp.14, 28, 29)
Mar.	**28, 1941**	Plutonium 239 determined to undergo fission with slow neutrons by Seaborg, Kennedy, Wahl & Segre. Confirmed on May 18, 1941. (Seaborg A, pp.34, 41)
May	19, 1941	Lawrence phones Compton re possibility of plutonium bomb. Compton wires Vannevar Bush. (Seaborg A, p.41)
June	28, 1941	OSRD established. (Nichols, p.33 and Groves p.127)
Nov.,	1941	Uranium project (S-1 Committee) placed under OSRD. (Groves, p.8)
Dec.	**6, 1941**	University of Chicago Met Lab plutonium program begins. (Groves p.9 and Jones, p.33 and Seaborg A, p.v.)
First	Qu. 1942	Met Lab ships uranium chloride to UC for transformation to plutonium chloride in UC's cyclotron. UC ships plutonium chloride to Met Lab. (Nelson)
Mar.	**9, 1942**	Vannevar Bush recommends Corps of Engineers be given charge of construction. (Jones, p.37)

Mar.	19, 1942	Seaborg suggests the name "Plutonium" for element 94 in report No. A-135 of this date. (Seaborg A, pp.89-91)
May	14, 1942	Conant reports on the five most promising production and separation methods, and surmises Germany might be in the lead. (Smyth, p.80)
June	18, 1942	Corps is assigned responsibility for the bomb project. (Groves, p.10 and Smyth, p.83)
June,	29, 1942	Corps engages Stone & Webster as overall contractor for both Chicago and Clinton work. (Nichols, p.36, 37 and Groves, p.13)
July	24, 1942	Seaborg first writes about bismuth phosphate as a possible carrier for plutonium. (Seaborg A, p.172)
Aug.	3, 1942	DuPont's Charles Cooper reports for work at Met Lab to begin work on chemical engineering of the separation process. (Seaborg A, p.173)
Aug.	13, 1942	Corps selects "Manhattan Engineer District" (MED) as designation. (Nichols, p.40)
Aug.	20, 1942	Met Lab produces first visible ^{239}Pu compound, a fluoride. (Seaborg A, p.176)
Sept.	13, 1942 and 14	S-1 meeting of physicists at Bohemian Grove on the Russian River, California. Agreement reached on scope and on cooperation with the MED. (Nichols, p.44 and Jones, p.71 and Seaborg, pp.185, 187)
Sept.	17, 1942	Groves appointed Chief of MED.) Groves, p.417 and (Seaborg A, p. 191)
Sept.	19, 1942	Groves obtains AAA priority from Nelson, WPB chief. (Groves, p. 22)
Sept.	23, 1942	By the President's direction, a meeting was held between the Secretary of War and the Army Chief of Staff re: project priorities. A new Military Policy Committee was formed, with Groves named Executive Secretary, thus placing all phases of the Manhattan program under Groves. (Jones, p.77)
Sept.	28, 1942	Corps requests DuPont to assist Stone & Webster in design and procurement of equipment for Chicago separation plant (DuPont A, p.10 and DeRight, p.1)
Oct.	3, 1942	DuPont and Corps agree on letter contract for design and procurement of equipment for Chicago semiworks separation plant. Not executed until Oct. 27. DuPont however, immediately organizes plutonium-process design group (DuPont A, pp.11, 173 and DeRight, p.1)
Oct.	5, 1942	Ray Genereaux begins separation-plant engineering (Genereaux M)
Oct.	6, 1942	DuPont Supervising Engineer makes scoping visit to Met Lab to review entire program. (DuPont A, p.11)
Oct.,	1942	DuPont considers periscopes and TV for monitoring separation process. (DuPont A, p.189)
Oct.	10, 1942	Corps asks DuPont to supplement their separation process work by designing and procuring equipment for the pile. DuPont refuses. (DuPont A, p. 11 and Jones, p.101 and DeRight, p.1)

Oct.	30, 1942	Groves & Conant in Washington, D.C. request DuPont's Stine & Harrington opinion on feasibility of (a) DuPont's beginning immediately the design of full-scale plutonium production plant and (b) the successful operation of this plant in a reasonable time. Groves and Conant express grave doubts about feasibility of this process. DuPont says they are not qualified to comment on operational feasibility and will refer design question to their Executive Committee (DuPont A, p.14 and Groves, pp.46, 47)
Nov.	4-6, 1942	Following meeting of DuPont Executive Comm., DuPont's Stine, Bolton, Williams, Gary, Chilton, Greenwalt, Pardee, and Johnson visit Met Lab to review program status. (DuPont A, p.15 and DeRight, p.2 and Nichols, p.63 and Seaborg A, p.202)
Nov.	6, 1942	First Met Lab studies of bismuth phosphate carrier. (Seaborg A, p.204)
Nov.	10, 1942	DuPont requests review of alternative processes leading to the bomb. (DuPont A, p.16 and Groves, p.51)
Nov.	**12, 1942**	Agreement by Walter Carpenter and the Executive Comittee to undertake the full-scale plutonium production facility. (Groves, p.51)
Nov.	14, 1942	Fermi and Compton decide to relocate Pile No. 1 to West Stands because of labor problems at Argonne Forest. (Seaborg A, p.207)
Nov.	16, 1942	Met Lab separation-process design in progress under Cooper; wet and dry fluoride and wet peroxide. This design for Hanford plant. DuPont A, p.177 and Seaborg A, p.208)
Nov.	18, 1942	Groves appoints Lewis Committee – W.K. Lewis, Tom Gary, Craword Greenwalt, Roger Williams, Eger Murphree – in response to DuPont's request for alternatives review. (DeRight, p.2 and Nichols, p.65 and DuPont A, p.17)
Nov.	19, 1942	Separation-building crane-design conference; DuPont and Whiting Corp. (DuPont A, p.188)
Nov.	26, 1942	Lewis Committee first visits Met Lab to hear their presentation. (Compton, p.135 and Seaborg A, p.215)
Dec.	**1, 1942**	Superseding Corps/DuPont letter contract prepared for design and construction of full-scale plutonium-production plant at Clinton. Corps later accepts DuPont recommendation to relocate this plant to a more remote site. DuPont agreed to this contract on Dec. 21 and executed it on Dec. 23. (DeRight, pp.2, 3 and DuPont A, p.21)
Dec.	**2, 1942**	First sustained nuclear chain reaction in history under West Stands. Pile No. 1 goes critical about 3:25 PM while Lewis Committee is making its second visit to Met Lab. (Seaborg A, p.217, and Fig.5.1)
Dec.	**4, 1942**	Lewis committee final report: Diffusion most likely to produce large quantities, but recommend construction of magnetic and plutonium-pile pilot plants, plus several full-scale piles. Groves receives on Dec. 7. (DuPont A, p.17 and Jones p.104)
Dec.,	1942	Met Lab sends helium-based design description to Wilmington. DuPont begins preliminary helium design; it runs into Jan., 1943. Shielding thicknesses calculated. (DuPont A, pp.61, 73)

Dec.	7. 1942	DuPont provides initial helium design. (DuPont A, p.73)
Dec.,	1942	DuPont instructed to design four 250,000 kW helium-cooled piles. DuPont prepares list of questions on this concept. (DuPont A, p.74)
Dec.	10, 1942	Military Policy Committee determines necessity for a new and remote site for plutonium production (Jones, p.109)
Dec.	**14, 1942**	Site-requirements meeting in Wilmington – Corps – Met Lab – DuPont. Intra plant and public clearances. Permanent town clearances. Townsite narrowed to either Benton City or Richland if site turns out to be Hanford. (App.H-1, 1.FM and Groves, p.70 and DuPont B, p.0005191)
Dec.	15, 1942	Col. Matthias and DuPont's Gil Church and Al Hall review the Corps file data in Washington, D.C on the candidate sites. (App. H-1, 3.FM)
Dec.	17, 1942	Two-day Met Lab/DuPont design meeting in Wilmington to consider DuPont's list of questions on the 250,000 kW piles. DuPont instructed to design this helium-cooled plant. (DuPont A, p.74)
Dec. 18-30, 1942 *		Site team arrives Seattle Dec. 18. Meetings with District personnel there and at Spokane, Portland, Sacramento, and Los Angeles. On-site inspections in Idaho, E. Washington including Hanford, and Blythe, CA. Overflights of north-central Oregon, Horse Heaven, Needles, CA, and Blythe sites. (App.H-1, 3-15.FM and Matthias A, Dec. 29)
Dec.	28, 1942	Corps and DuPont meet National Carbon re: scoping, max. block size, tolerances and graphite manufacturing methods. (DuPont A, p.95)
Dec.	30, 1942	DuPont and National Carbon discuss block size and dimensional proportion limitations, machining factors and methods, purity/size relationship, lattice dimensions, and pile temperature limitations. (DuPont A, P.95)
Dec.	30,*1942	Site team departs LA; arrives Washington, D.C. Dec. 31. (App. H-1, 12.FM)
Dec.	31, 1942 P	Corps requests graphite estimate from DuPont. This was 1.5 months before final decision for a water-cooled, cubical pile; helium design still in progress. (DuPont A p.349)
Jan.	**1, 1943**	Corps site-decision meeting at HQ; Groves, General Thomas Robbins, Matthias, Carl Giroux. Hanford chosen, pending Groves trip and Met Lab meteorological decision. Coulee power a deciding factor. (App.H-1, 16.FM)
Jan.,	1943 P	DuPont prepares preliminary procurement schedule of critical items: centrifuges, cranes, valves, pumps, motors, heat exchangers, transformers, etc. Target procurement dates established. (DuPont A, p.307)
Jan.,	1943 P	DuPont meets Alcoa to discuss fuel-tube manufacturing feasibility. Order placed for development of extrusion method. (DuPont A, p.353)

Jan.,	1943	Preliminary site layout complete. This had been drafted by the site team before they left Washington, D.C. in December. (App.H-1, 17.FM)
Jan.	4, 1943	DuPont awarded design of Clinton, Tenn. semiworks by MED. (DuPont A, p.173)
Jan.	9, 1943	Met Lab (Wigner) recommends study of water-cooling alternative to DuPont who agree and begin water-cooling design. Helium design continues as first priority, however. (DuPont A, p.76)
Jan.	9, 1943	Matthias briefs Col. Nichols on Hanford in New York City. (App. H-1, 16.FM)
Jan.	11, 1943 *	Matthias and DuPont fly to Spokane and drive to Hanford to brief Portland District real-estate people. (App.H-1, 16.FM)
Jan.	14, 1943 P	DuPont forwards graphite tonnage estimate to Corp. P (DuPont A, p.349)
Jan. 14-19, 1943 *		Groves and Matthias inspect Hanford and vicinity. (App.H-1 18.FM)
Jan.	20, 1943	Met Lab sends DuPont Report No. 407, dated 1/9 for a cylindrical, water-cooled pile. (DuPont A, pp.76, 77)
Jan.	22, 1943 P	DuPont discusses helium compressors with Ingersoll Rand. (DuPont A, p.75)
Jan.	23, 1943	Groves and DuPont in Wilmington select Richland as permanent townsite. (DuPont B, p.0005191)
Feb.	3, 1943	Seaborg sends Greenwalt summaries of six alternative separation processes. (Seaborg A, p.241 and DuPont A, p.174)
Feb.,	1943 P	DuPont places uranium billet extrusion order with B&T Metals. (DuPont A, p.366)
Feb.,	**1943**	Matthias appointed as Oficer In Charge at Hanford. (App. H-1, 20.FM)
Feb.,	8, 1943	Undersecretary of War Patterson approves Hanford land acquisition request. (Groves, p.75. Jones, p.331 says Feb.9)
Feb.	10, 1943	DuPont and Met Lab discuss machining of graphite blocks, including test methods, graphite properties, tool types, and tool maintenance. (DuPont A, p.99)
Feb.	13, 1943	DuPont and Met Lab meet re: helium vs. water cooling. Four principle design tasks assigned DuPont. (DuPont A, pp.78, 79)
Feb.	**16, 1943**	DuPont and Met Lab agree on water-cooled cubical pile as final design. (DuPont A, p.79 and DeRight, p.6)
Feb. 16-17, 1943		DuPont, Speer Carbon, and International Graphite meetings concerning machinery types. (DuPont A, p.100)
Feb.	17, 1943	Met Lab now reports to Groves rather than to S-1. (Seaborg A, p. 249)
Feb.	22, 1943	DuPont/National Carbon conference on graphite blocks: cored vs. bored bars as related to dimensions. (DuPont A, p.97)
Feb.	25, 1943 *	DuPont arrives Pasco to arrange temporary office space for Corps. (DuPont B, v.1, p.l6)

Feb.	25, 1943	Met Lab gives final approval for Hanford site, as governed by meteorological requirements. (DuPont A, p.19)
Mar.,	1943 *	Matthias moves to Pasco. DuPont leases 18 buildings in Pasco. Camp construction starts. (App. H-1, 27, 49.FM and DuPont B, v.l)
Mar.,	1943 *	100-B staked for general site excavation. (Gerber, p.9) Matthias said that excavation started in "early April", which is not incompatible w/Gerber. (App.H-1, 52.FM)
Mar.,	1943	Separation-cell final design complete. (Genereaux F)
Mar.	3, 1943	Slug-canning efforts begin. (DuPont A, p.384)
Mar.	17, 1943	Decision for 2 separation areas – East & West – 4 miles apart; 3 separation plants each. (DuPont A, p.178)
Mar.	**22, 1943 ***	Official start of construction; opening of DuPont's employment office in Pasco. (Hageman, p.39 and Title Sheet, DuPont A)
Mar.	23, 1943 *	Gil Church, DuPont's construction manager moves to Pasco. (DuPont B, v.1, p.16)
Mar.	26, 1943 *	First contact with Milwaukee RR. (Matthias A, Mar. 26)
Mar.	29, 1943	Bismuth phosphate process shows great promise. (Seaborg A, p. 258)
Mar.	31, 1943 P	First extrusion of uranium rods by B&T Metals, Columbus, Ohio. (DuPont A, p.367)
Apr.	6, 1943 P	Final selection of 4-3/16 in. square graphite blocks by 48 ins. long (coordinated with final 8-3/8 in. lattice. (DuPont A, p.350)
Apr.	23, 1943	TNX specifies iron and cellulose thicknesses for the biological-shield blocks. (DuPont A, p.107)
Apr.	25, 9943 *	Initial occupation of camp housing (tents). (Matthias A, Apr. 25)
May,	1943	Decision to build 4 separation buildings, not 6. (DuPont A, p.179)
May,	1943 P	First production purchase order for slug machining. (DuPont A, p.380)
May	11, 1943	DuPont approves CMX construction. (DuPont A, p.85)
May	27, 1943	Decision to build 3 reactors, not 4. (Matthias A, May 27)
June,	1943 P	Stainless steel procurement begins. Plate bought and stored for shipment to subsequent fabricators. (DuPont A, p.309-313)
June,	1943	Design start for CMX. (DuPont A, p.85)
June,	1943	Thermal shield determined necessary. (DuPont A, p.85)
June	**1, 1943**	Separation-selection conference at Met Lab. Greenwalt chooses bismuth phosphate process for high-radiation end of process. Seaborg A, pp.279-283)
June	8, 1943	Final design at Wilmington re-directed to bismuth phosphate. (DuPont A, p.175)
June	10, 1943 *	Preliminary construction starts at 100 B (Matthias A, June 10)
June	12, 1943 P	Corps instructs DuPont to bypass 3-bid requirement for steel plate upon receipt of WPB allocation, or slot in rolling schedule. (DuPont A, P.308)

June	22, 1943 *	200 West Cell building excavation start. (DuPont B Progress Charts)
Aug.,	1943 P	20,000 tons of steel plate ordered for biological shield blocks. (DuPont A, p.323)
Aug.,	1943 P	Shield block fabricating orders placed. (DuPont A, p.330)
Aug.,	1943	SMX experiments begin at Clinton.DuPont A, p.83)
Aug.,	1943	Control-rod final design start. (DuPont A, p.117)
Aug.	13, 1943	Met Lab makes successful runs on extraction step. (Seaborg A, p.299)
Aug.	5, 1943 P	7,500,000 sf. of special Masonite sheet ordered for biological shield blocks. (DuPont A, p.326)
Sept.,	1943	Biological shield block dimensions finalized, and first drawing released. (DuPont A, pp.108,330)
Sept.,	1943 P	Uranium slug extrusion and machining problem solved. (DuPont A, p.381)
Sept.	1, 1943 *	CMX corrosion tests begin at Hanford. (DuPont A, p.85)
Oct.,	1943	Pile specifications complete. (Jones, p.193)
Oct.	10, 1943 *	105 B building excavation start. (DuPont B, progress charts)
Nov.	1, 1943	Decision to establish Wilmington Shops experimental line for canning Hanford slugs. (DuPont A, p.390)
Nov.	**4, 1943**	Clinton pilot pile begins operation. (Seaborg A, p.348)
Nov.	**6, 1943**	Definitive contract No. W-7412-Eng-1 signed; Corps & DuPont. (DuPont A, p.21)
Dec.	10, 1943	Machining of graphite blocks for B Reactor starts at 101 Building. (DuPont B, p.790)
Dec.	14, 1943 P	Alcoa proves out final method of drawing aluminum fuel tubes – production starts. (DuPont A, p.356)
Dec.	**19, 1943**	Separation process test run at Clinton proves out bismuth-phosphate process. (DuPont A, p.176. Seaborg A says Dec.20, p.34)
Jan.	**1, 1944**	Clinton sends first milligram quantities of plutonium to Met Lab, who then begin confirmation of Pu chemistry with the metal, and determination of metallurgical properties. (Seaborg A, p.365)
Jan.	2, 1944	First shield block for Reactor shipped. (Fig.5.7)
Jan.	23, 1944 *	Separation cell building – West – foundations complete. (DuPont B progres chart)
Jan. ,	1944	Final decision by Met Lab, Grasseli labs, & TNX on type of slug canning. Hanford 300 Area though, still had 7 months of development of the actual method left to do. (DuPont A, p.391 and App.H-2, 10 , 11.WS)
Jan.–Aug., 1944		Extreme difficulties in extruding uranium in 2% of the rods for Hanford. The Corps, Battelle, Met Lab, and TNX determine high iron content as cause. Problem worked out by Aug. (DuPont A, pp.372-371)

Feb.,	1944	Slug-discharge end of reactor: final design starts. Decision for full-scale model at Wilmington Shops of the discharge canal. (DuPont A, p.125)
Feb.	1, 1944	First detailed separation-process flow sheet, although several months of development work were required for completion. (DuPont A, p.176)
Feb.	4, 1944	Biological-shield tie strap conference in Wilmington Design Division. Decision to consult welding experts. (DuPont A, p.109)
Feb.	7, 1944	Welding conference at Wilmington – DuPont, Lukens Steel, Combustion Engineering, New York Shipbuilding. (DuPont A, p.110)
Feb.	17, 1944 *	F Reactor construction start. (DuPont B progress chart)
Mar.,	1944	Separation process chemistry essentially complete, although final details of chemical engineering will not be worked out until September. (Seaborg A, pp.421, 422
Mar.	31, 1944 *	105 B Building foundations complete. (DuPont B progress chart)
Apr.,	1944	Major redesign of Purification bldg. 231 required due to relaxation of criticality requirement. (DuPont A, p.203)
Apr.	21, 1944	HEW drawings (5700 total) 80% complete. (DuPont B Design Status Charts.)
Apr.	26, 1944	Start of design of shielded discharge cab. (DuPont A, p.130)
May	17, 1944 *	Start of machining of graphite blocks for D Reactor at 300 Area. (DuPont B, v.3, p.790)
June	1, 1944 *	B Reactor pile layup complete. (DuPont B, v.3 p.790)
June	**20, 1944**	Seaborg's chemistry group achieved its goal of a decontamination factor of 5×10^5 in the bismuth phosphate process, and has modified it to handle the low-plutonium concentrations of the initial Hanford production. (Seaborg A, p.471)
June	28, 1944 *	D Reactor pile layup complete. (DuPont B, v.3 p.790)
July	1944 *	Slug-canning process at 300 Area achieves adequate percentage of leak-proof canned slugs. (DuPont A, p.394 and App.H-2, 10, 11.WS)
July	1, 1944	Decision for three alternative mechanical methods of isolation of product in 231 building. (DuPont A, p.203)
July	21, 1944 *	105 B building superstructure complete. (DuPont B, Progress Chart)
Aug.	, 1944 P	Uranium rod extrusion complete. (DuPont A, p.369)
Aug.	19, 1944 *	B Reactor equipment installation complete, but unfueled, (DuPont B, Progress Chart)
Sept.	8, 1944 *	Separation cell building superstructure complete. (DuPont B, Progress Chart)
Sept.	12, 1944	Determination of need to subsitute air jets for steam jets in 231 building to avoid peroxide explosion. (DuPont A, p.204)
Sept.	**13, 1944** *	B Reactor fueling begins. (Mathias A, Sept. 13 and Seaborg A, p.19 and Jones p.219)

Sept.	15, 1944 *	Separation cell building – West – equipment installation complete. (DuPont B, progress chart)
Sept.	**26, 1944** *	B Reactor goes critical. Xenon poisoning occurs. (Jones, p.221 and Gerber, p.9. Seaborg A, p.529 says "a few minutes after midnight" in the early AM of the 27th)
Sept.	27, 1944 *	Separation cell building – West – complete. (DuPont B, progress chart)
Oct.	30, 1944 *	CMX tests at Hanford conclude. Result: the $10 million demineralizer and the B, D, and F deaerators were not required. (DuPont A, p.86)
Nov.	10, 1944	Reactor – 100 Area – drawings 100% complete. (DuPont B, Design Status Charts)
Nov.	18, 1944 *	F Reactor pile layup complete. (DuPont B, v.3, p.790)
Nov.	**22, 1944**	Met Lab input to Hanford design and operations is complete. (Seaborg A, p.563)
Nov.	24, 1944 *	First discharge of irradiated slugs from B Reactor (from partial loading). (Gerber A, p.9)
Dec.	**6, 1944** *	200 Area operational start. (T Plant). Seaborg A, p.581 and Gerber B, p.4 and DuPont B Building and Facilities Charts.
Dec.	17, 1944 *	D Reactor start (Seaborg A, p.589 and Rhodes, p.560)
Dec.	**28, 1944** *	B Reactor re-start. (Matthias A, Dec. 29 and Rhodes, p.560)
Feb.	**5, 1945** *	First plutonium nitrate shipment to Los Alamos; hand carried by Matthias to Los Alamos messenger at Los Angeles. (App.H-l, 59.FM)
Feb.	25, 1945 *	F Reactor startup. (Seaborg A, p.629)
June	1, 1945	S-1 Committee's final recommendation to Truman to use the bomb. (McCullough B, pp.390, 391)
June	**15, 1945** *	Hanford had shipped enough plutonium for the Trinity bomb and one full-sized bomb. (App.H-2, 16.WS)
July	**16, 1945**	Trinity test of plutonium implosion-type bomb Alamagordo, New Mexico. (Groves p.293)
Aug.	6, 1945	Uranium, gun-type bomb dropped. (Groves, p.318)
Aug.	9, 1945	Plutonium, implosion-type bomb dropped. (Groves, p.342)
Aug.	14, 1945	Japanese surrender. (Seaborg A, p.746)
Aug.	18, 1945	A third bomb – plutonium implosion-type – was scheduled for a drop on this date but was unnecessary. (Nichols, p.215, quoting Groves Aug. 10 memo to General Geoge Marshall)
Dec.	, 1945	The second uranium, gun-type bomb was ready. (Hewlett and Anderson, p.334)

Appendix G

SUPPORTING CALCULATIONS

Calculation No.	Subject	Page No.
1.	HEW Cost Escalation to 1994 From 1945	143
2.	DuPont's Management & Engineering Costs	149
3.	Determine The Following Plutonium-Project Schedules: Actual Wartime; Rational Wartime; and Rational Peacetime	151
4.	Determine Numerical Comparisons Among Magnitude Data of Chapter 2	157

SUPPORTING CALCULATION
FOR
REPORT ON MANAGEMENT OF THE HANFORD ENGINEER WORKS IN WW II

CALCULATION SUBJECT: HEW COST ESCALATION TO 1994 FROM 1945

DATE: Sept. 28, 1995	CALCULATION NUMBER 1	Page 1 of 6

OBJECTIVE. Obtain the following 1994 costs:

o Total cost, with and without first uranium charge, both without profit.
o 1994 profit
o Total cost, with and without first uranium charge, but with profit added to both
o Total cost of 200 Area, including distributed indirects and profit.
o 1994 procurement values: total Wilmington, total Hanford, and total HEW
o Total Manhattan project

INTRODUCTION

The escalation was done using the following increase factors from Table 7.2:

o Average of Chem. Eng. Plant and Nelson-Farrar factors: 10
o ENR Construction-Cost index   17.7
o Consumer-Price index   8.2

These are applied to the equipment-related, construction, and salary-related line-item costs of Table 7.1, as detailed on p.3.

The only real-world, 1994 adjustment I made was to add profit, because DuPont did the original job for a nominal $1 profit. I calculated the profit by the method given in Department of Energy Acquisition Regulation No. DOE/MA-0189, June 1985 (the "DEAR" method).

Making further adjustments based on today's conditions would not have meant much. In the first place, you probably couldn't have built on the Hanford site because of a population increase since 1945, and for other reasons.

The 1994 construction camp, worth $252 million in 1994 dollars, would have been much smaller and less costly because there would have been no wartime gas shortage; workers would have commuted from 50 miles away, or more. On the other hand, the project would have lasted 5 or 6 years (See Figure 5.5) because they wouldn't have worked overtime. The added length of the job would have increased administrative and overhead costs.

There would have been many other adjustments between the conditions of 1945 and 1994, so the only reasonable thing to do was to escalate by indices, plus the addition of profit.

Conclusions, Given Data, and Assumptions are provided on next page.

SUPPORTING CALCULATION
FOR
REPORT ON MANAGEMENT OF THE HANFORD ENGINEER WORKS IN WW II

CALCULATION SUBJECT: HEW COST ESCALATION TO 1994 FROM 1945
DATE: Sept. 28, 1995 CALCULATION NUMBER   1   Page 2 of 6

CONCLUSIONS

    HEW total cost without profit, without first U charge: $4.111 bil.
                Ditto   with first U charge: $4.351 bil.
    1994 profit $0.018 bil.

    HEW total cost with profit, without first U charge $4.129 bil.
                Ditto   with first U charge $4.369 bil.

    Total cost of 200 Area, including distributed $0.775 bil.
      indirects and profit

    1994 procurement values:
      Wilmington: $0.654 bil.
      Hanford: $2.184 bil.
      Total HEW Procurement: $2.838 bil.

    Escalated total Manhattan Project, 1994: $26.9 bil.

GIVEN DATA

    Recorded costs as given in Table 7.1
    Cost increase factors as given in Table 7.2
    Procurement data as shown in Item 4, p.5 of this calculation
    Total cost of the Manhattan Project was $2.2 billion, 1945.

ASSUMPTIONS

    1994 profit would have been 80% of the maximum DEAR profit.
    Total Manhattan-Project cost may be escalated by the ratio
      between the escalated 1994 Hanford cost and the 1945 Hanford cost.

SUPPORTING CALCULATION
FOR
REPORT ON MANAGEMENT OF THE HANFORD ENGINEER WORKS IN WW II

CALCULATION SUBJECT: HEW COST ESCALATION TO 1994 FROM 1945

DATE: Sept. 28, 1995   CALCULATION NUMBER   1   Page 3 of 6

1. Escalation Of Line-Item Costs Of Table 7.1

	Table 7.1 Costs	CH.EN NEL-F @ 10	ENR @ 17.7	CPI @ 8.2	Escalated Cost, $Million
MANUFACTURING					
100 Areas	$92,791,269	o			$928
200 Areas	53,750,693	o			538
300 Areas	7,122,992	o			71
500 Outside Electric Lines	3,390,878		o		60
600 Roads, Railroads, etc.	19,391,692		o		343
700 Administrative, Maintenance Area	3,086,323		o		55
800 Outside Overhead Pipe Lines	1,656,573		o		29
900 Outside Underground Pipe Lines	10,709,128		o		190
Spare Parts	--				
Site Work And General Grading	2,004,898		o		35
RICHLAND VILLAGE	34,834,286		o		617
Sub Total: Manufacturing Plus Richland Village:					$2866 Mil.
CONSTRUCTION FACILITIES					
Administrative Area - Hanford	2,491,865		o		44
Construction Camp - Hanford	14,237,045		o		252
Steam, Water, Sewers - Hanford	11,651,477		o		206
Camp Operation - Hanford	18,368,973			o	151
Temporary Construction Equipment	6,767,741	o			68
Major Construction Equipment	3,518,342	o			35
Small Tools	291,725	o			3
Handling Excess Materials	2,343,291			o	19
Excess Materials	237,513		o		4
COMMERCIAL FACILITIES (Bldg. & Equip.)					
Hanford Area	1,470,797		o		26
General, Including Alterations To Existing Houses	627,409		o		11
Camp 300 Area-Military Police	490,400			o	4
FIELD EXPENSE					
Fire and Police Protection and Sanitation	4,531,007			o	37
Safety and Medical	2,350,270			o	19
Recruiting and Incentive Plan	7,294,270			o	60
Supplies, Light, Heat, Water, etc.	3,238,150		o		57
Compensation Insurance	393,493			o	3
P.O.A.B. and Unemployment Tax	5,904,486			o	48
General Items	4,215,069		o		75
Fees-Pipe and Electrical CPFF Contractors	474,700		o		8
FIELD SUPERVISION	8,963,101			o	73
WILMINGTON PROCUREMENT, INSPECTION, AND SUPERVISION	1,819,838			o	15
DESIGN ENGINEERING	3,280,062			o	27

TOTAL 1994 ESCALATED HANFORD COST WITHOUT FIRST URANIUM CHARGE	$4111
FIRST URANIUM CHARGE: $24 MILLION(1945), ESCALATED 10 X (DuPont A, p.434)	240
TOTAL ESCALATED HANFORD COST WITH FIRST URANIUM CHARGE, w/o profit	$4351

SUPPORTING CALCULATION
FOR
REPORT ON MANAGEMENT OF THE HANFORD ENGINEER WORKS IN WW II

CALCULATION SUBJECT: HEW COST ESCALATION TO 1994 FROM 1945

DATE: Sept. 28, 1995 CALCULATION NUMBER 1 Page 4 of 6

2. <u>Hanford Profit on Escalated 1994 Cost</u>

In 1994 DOE would have allowed a small profit on the Hanford project. This profit would have been calculated by the DEAR regulation. The fee schedule from that regulation is included on pg. 6 of this calculation.

DOE would have allowed this fee on the basis of "Total Expenditures" By DuPont, as shown in Table 7.1.

Ratio 1994's $4.111 billion down to the "Total Expenditure" basis:

$\frac{308.7}{333.7}$ x $4.111 = $3.803 billion "Total Expenditures"
less $-.5$ billion, maximum tabular job cost (p.6)
$3.303 billion, basis for 0.52% fee increase

$3.303 x .0052 = $.0172
       + .0047 tabular profit (p.6)
         $.0219 Total nominal DEAR profit, 1994

This is only <u>nominal</u> profit. On June 27, 1994, DOE's procurement analyst for fee considerations, Mike Righi, said that this would be the <u>maximum possible</u> profit. The actual profit awarded by DOE depends on DOE's analysis of all project conditions. Mr. Righi said he has <u>never</u> seen the maximum awarded.

I therefore assume 80% x $.0219 billion, or a 1994 HEW profit of $.018 bil. Add to HEW totals w/ & w/o U charge, to obtain:
With U charge: $4.37 bil. Without U. charge: $4.13 bil.

3. <u>Distribution Of 1994 Profit And Escalated Indirects To Process-Area Direct Costs.</u>

As input to a cost statement I make in the Checking section of Chapter 5, it is necessary to find the total cost of the 200 Area. I calculate the 100 and 300 Areas, as well.

From p.3, the escalated indirects are: $Million
   Construction Facilities   782
   Commercial Facilities   41
   Field Expense   307
   Field Supervision   73
   Procurement, Inspection, and Wilmington Supervision $15
   Design Engineering   27
   Total Escalated Indirects $ 1245 mil.

Total Direct Cost, Escalated: Manufacturing $ 2866 mil.
   Plus Richland Village   (p.3)

(3. Cont. on next pg.)

SUPPORTING CALCULATION
FOR
REPORT ON MANAGEMENT OF THE HANFORD ENGINEER WORKS IN WW II

CALCULATION SUBJECT: HEW COST ESCALATION TO 1994 FROM 1945

DATE: Sept. 28, 1995   CALCULATION NUMBER   1   Page 5 of 6

3. (Cont.)

   100 Area: $\frac{928}{2866} \times (1245 + 18 = 1263) = 409$
   $+ 928$
   $\$1337$ mil.

   (Indir.+Prof.)

   200 Area: $\frac{538}{2866} \times 1263 = 237$
   $+ 538$
   $\$ 775$ mil.

   300 Area: $\frac{71}{2866} \times 1263 = 31$
   $+ 71$
   $\$ 102$ mil.

4. Escalate Purchase-Order Value

   For Wilmington orders - largely the process equipment - use the average of Nelson-Farrar/Chem. Eng., 10. For Hanford orders - conventional equipment, materials, and subcontracts, use the ENR, 17.7.

   1945 purchase order values are from DuPont B, pp.223-225.
   1945 subcontract values are from DuPont A, pp. 292,293.

   Wilmington: $65,430,000 x 10, rounded $654,000,000

   Hanford: $81,930,000 x 17.7 $1,450,000,000
   Subcontracts: $65,890,000 x 17.7   1,166,000,000
   Total Hanford: $2,616,000,000   2,616,000,000

   Total Procurement: $3,270,000,000

5. Escalate Total Manhattan-Project Cost

   Assume that Manhattan-Project escalation is proportional to the above-determined HEW increase.

   1945 HEW cost with first uranium charge was $333.7 mil. plus $24 mil. = $357.7 mil.
   1994 HEW cost with first U charge & profit was $4.37 bil.

   Then: $\frac{\$4.37 \text{ bil.}}{\$0.3577 \text{ bil.}} \times \$2.2 = \$26.9$ bil., escalated Manhattan cost.

SUPPORTING CALCULATION
FOR
REPORT ON MANAGEMENT OF THE HANFORD ENGINEER WORKS IN WW II

CALCULATION SUBJECT: HEW COST ESCALATION TO 1994 FROM 1945

DATE: Sept. 28, 1995 **CALCULATION NUMBER** 1 Page 6 of 6

5. DEAR FEE
   SCHEDULE.

915.971-5

AMENDMENT 13
DEPARTMENT OF

(c) The fee schedule shown in paragraphs (d) and (f) assumes a letter of credit financing arrangement. If a contract provides for or requires the contractor to make their own cost investment for contract performance (i.e., when there is no letter-of-credit financing), the fee amounts set forth in the fee schedules shall be increased by an amount equal to 5 percent of the fee amount as determined from the schedules.

(d) The following schedule sets forth the base for construction contracts: Fee's profit (1)

CONSTRUCTION CONTRACTS SCHEDULE

FEE BASE (dollars)	FEE (dollars)	FEE (%)	INCR (%)
100,000	5,400	5.40	5.30
300,000	16,000	5.33	5.00
500,000	26,000	5.20	4.80
1,000,000	50,000	5.00	3.55
3,000,000	121,000	4.03	3.00
5,000,000	181,000	3.62	2.62
10,000,000	312,000	3.12	2.38
15,000,000	431,000	2.87	2.01
25,000,000	632,000	2.53	1.79
40,000,000	900,000	2.25	1.58
60,000,000	1,216,000	2.03	1.43
80,000,000	1,502,000	1.88	1.29
100,000,000	1,759,000	1.76	1.15
150,000,000	2,335,000	1.56	0.99
200,000,000	2,829,000	1.41	0.73
300,000,000	3,563,000	1.19	0.65
400,000,000	4,186,000	1.05	0.52
500,000,000	4,706,000	0.94	
Over $500 million	4,706,000		0.52*

* 0.52% excess over $500 million.

(e) When using the Construction Contracts Schedule for establishing maximum payable basic fees, the following adjustments shall be made to the Schedule fee amounts for (1) complexity levels, (2) excessive subcontracting, (3) normal contractor services performed by the government or another contractor:

(i) The target fee amounts, set forth in the fee schedule, shall not be adjusted for a Class A project, which is maximum complexity. A Class B project requires a 10 percent reduction in amounts. Class C and D projects require a 20 percent and 30 percent reduction, respectively. The various classes are defined in 915.971-4(b).

(ii) The target fee schedule provides for 45 percent of the contract work to be subcontracted for such things as electrical and other specialties. Excessive subcontracting results when such efforts exceed 45 percent of the total contract work. To establish appropriate fee reductions for excessive subcontracting, the negotiating official should first determine the amount of subcontracting as a percentage of the total contract work. Next, the negotiating official should determine a percentage by which the prime contractor's normal requirement (based on a requirement for doing work with its own forces) is reduced due to the

(1) DOE I.G. office, ...
     SAN
915-20   6-15-64 ...

SUPPORTING CALCULATION
FOR
REPORT ON MANAGEMENT OF THE HANFORD ENGINEER WORKS IN WW II

CALCULATION SUBJECT:	DUPONT'S MANAGEMENT & ENGINEERING(M&E) COSTS

DATE: Oct. 3, 1995	CALCULATION NUMBER 2	Page 1 of 2

OBJECTIVE. To determine if DuPont's management and engineering costs were reasonable, by comparison with ASCE recommendations.

INTRODUCTION.

1. DuPont's costs for Wilmington management, design engineering, field supervision and total Capital Investment are given in Table 7.1.

2. Adjustments to the M&E costs were necessary to permit comparison with ASCE recommendations. The adjustments were for wage increases, which did not occur during WW II, and for profit, which DuPont declined.

3. Adjusted costs were then compared with ASCE recommendations contained in their Manual 45.

CONCLUSIONS.

1. DuPont's management and engineering costs were not only not excessive, they were at and below the low end of ASCE recommendations.

GIVEN DATA. The recorded costs of Table 7.1.
Recommendations of ASCE Manual 45, 1968 & 1988 editions.

ASSUMPTIONS.
1. Upward adjustment of wages in accordance with the CPI
2. 85% of the M&E costs were wage-related.
3. Profit would have been 5% of DuPont's M&E costs.

1. Total Management and Engineering Costs.

    These are made up of the following line-item costs from Table 7.1 (to the nearest $1000)

Field Supervision	$8,963,000
Wilmington procurement, inspection, & supervision	$1,820,000
Design engineering	+ $3,280,000
Total Management and Engineering	$14,063,000

2. Upward Adjustments Necessary.

    2.1 Wage Rates were frozen during the war. I adjusted these upwards by the CPI (0.032% in 1943 and 0.043% in 1944).
    $1.032 \times 1.043 = 1.076$. Assuming that 85% of M&E is wage based:
    $0.85 \times .076 = 0.0646$

SUPPORTING CALCULATION
FOR
REPORT ON MANAGEMENT OF THE HANFORD ENGINEER WORKS IN WW II

CALCULATION SUBJECT: DUPONT'S MANAGEMENT AND ENGINEERING COSTS

DATE: Oct. 3, 1995   CALCULATION NUMBER 2   Page 2 of 2

2.2 **Profit** must be added because DuPont did the job at cost. Because 1945 was long before the DEAR regulation was in effect assume that DuPont was entitled to 5% of their M&E costs.

2.3 **Total Adjustment** .05 + .0646 = .1146

From (1) above:

Total M&E x 1.1146 = $15,675,000
Des. Eng x 1.1146 =   3,656,000
Fld. Supv x 1.1146 =   9,990,000

3. **Comparison With ASCE Fee Recommendations.**

As recently as 1988 the ASCE published fee recommendations in proportion to the project's capital investment. (They currently recommend negotiations based on each project's individual conditions.)

It will be useful however, to compare DuPont's M&E expense with the former ASCE recommendations, as follows:

DuPont Activity	Adjusted DuPont Expense	Adjusted DuPont Expense as % of Total Capital Investment: $333,699,075	ASCE Recommended %
Total M&E	$15.68 mil.	4.7%	5.2% (1)
Design Engrng.	3.66 mil.	1.1%	4% - 9% (2)
Field Supervision	9.99 mil	3.0%	2.5% - 8% (2)

(1) ASCE Manual 45, 1968
(2) ASCE Manual 45, 1988

SUPPORTING CALCULATION
FOR
REPORT ON MANAGEMENT OF THE HANFORD ENGINEER WORKS IN WW II

CALCULATION SUBJECT: DETERMINE THE FOLLOWING PLUTONIUM-PROJECT
SCHEDULES: ACTUAL WARTIME; RATIONAL WARTIME; & RATIONAL PEACETIME.

DATE: July 23, 1994   CALCULATION NUMBER   3   Page 1 of 6

OBJECTIVE: Using the Actual, parallel Wartime Schedule as a base, calculate and draw the Rational, semi-sequential Wartime and Peacetime schedules.

INTRODUCTION: The plutonium project was completed by planning its activities in parallel: science; conceptual and preliminary design; semi-works; final design; procurement/construction; operations; and testing and dropping of the plutonium bomb.

The dates of these seven activities may be obtained from the record and may thus be plotted. The hours and shifts I determined from the interviews and by making certain assumptions.

It is possible then, to convert the parallel Actual Wartime Schedule to the Rational Wartime and Rational Peacetime Schedules by making some calculations and assumptions.

1. ACTUAL WARTIME SCHEDULE.   (Note: In source parentheses below, "S" equals Seaborg A.)

1.1 Science. Begins with the demonstration of the fissionability of ^{239}Pu on March 28, 1941 by Seaborg, Kennedy, Segre, and Wahl, and ends with the Clinton semiworks discovery of peroxide explosion with elevated temperature on Sept. 12, 1944 (pp.39,41 S. and p.204, DuPont A)

$$\begin{aligned}&3\text{-}28\text{-}41 \text{ to } 3\text{-}28\text{-}44 = 3 \text{ years}\\&\underline{3\text{-}28\text{-}44 \text{ to } 9\text{-}12\text{-}44 = 6 \text{ months}}, \text{ rounded}\\&\text{Science duration} = 3 \text{ years } 6 \text{ months}\end{aligned}$$

Working Hours: Tepe reported that the Met Lab scientists worked seven-day weeks. He said that their hours were very irregular and long. I made a conservative assumption of ten-hour days. (App. H-4,5.JT & 6.JT)

1.2 Conceptual/Preliminary Design. Begins with DuPont's appointment of a Supervising Engineer and a Design Project Manager on Oct. 3, 1942, and ends when DuPont and the Met Lab agree on a cubical, water-cooled, uranium slug, aluminum-tubed pile on Feb. 16, 1943. (pp.79,173 DuPont A).

Only average dates can be selected, as conceptual, preliminary, and final designs merged into each other. Final design of the separation cells, for example, began in October, 1942.

10-3-42 to 2-16-43 = 4 months, rounded

Working Hours. Genereaux reported that the Wilmington Design Division worked 5½ day weeks (44 hours). (App.H-3,14.RG)

SUPPORTING CALCULATION
FOR
REPORT ON MANAGEMENT OF THE HANFORD ENGINEER WORKS IN WW II

CALCULATION SUBJECT: DETERMINE PLUTONIUM-PROJECT SCHEDULES			
DATE: July 23, 1994	CALCULATION NUMBER 3	Page	2 of 6

1.3 **Semiworks.** Begins with Wilmington design upon receipt of Contract Supplement No. 1 on 1-4-43, and ends with the final input to Hanford design, the peroxide explosion information) on Sept. 12, 1944. (pp.81,204 DuPont A, and Exh.G,p.96 DuP.B)

  1-4-43 to 1-4-44 = 1 year
  1-4-44 to 9-12-44 = 8 months, rounded
  Semiw'ks duration = 1 year 8 months

Working Hours: Wilmington Design: 44 hour week, as in 1.2, above. Assume 5 months design. Clinton: Assume that construction and operations worked two 54 hr./week shifts.

1.4 **Final Design.** Assume begin on Feb. 16, 1943 (final pile scoping decision), and ends one month prior to the final 100-Area drawing - Oct. 10, 1994. I assume that the final month of drawings was occupied with final cleanup and revisions, and that design had ended a month earlier. (Design Status Graphs for 100-Areas, DuPont B)

  2-16-43 to 2-16-44 = 1 year
  2-16-44 to 10-10-44 = 8 months, rounded
  Final Des. Duration = 1 year 8 months, rounded

Working Hours: 44-hour work week, per 1.2, above.

1.5 **Procurement and Construction.** Begins with the canyon-crane meeting at Whiting Corp., Nov. 19, 1942, and ends at Hanford on Feb. 21, 1945. (p.188,DuPont A, and Matthias Diary entries of 2-2-45 and 2-21-45)

  11-19-42 to 11-19-44 = 2 years
  11-19-44 to 2-21-45 = 3 months, rounded
  Proc. & Constr. Dur. = 2 years 3 months rounded

1.6 **Operations.** Begins on Oct. 13, 1944 with the final loading of the 2004 tubes of the B Reactor and ends with production of sufficient plutonium for the Trinity and Nagasaki bombs on June 15, 1945. (Matthias Diary, Oct. 12, 1944, and App. H-2, 21.WS)

  10-13-44 to 6-15-45 = 8 months, rounded

Working Hours. Assume three 8-hour shifts.

SUPPORTING CALCULATION
FOR
REPORT ON MANAGEMENT OF THE HANFORD ENGINEER WORKS IN WW II

CALCULATION SUBJECT: DETERMINE PLUTONIUM-PROJECT SCHEDULES			
DATE: July 23, 1994	CALCULATION NUMBER 3	Page 3	of 6

1.7 Bomb Test And Drop. Begins with completion of plutonium shipments for Trinity and Nagasaki bombs to Los Alamos on June 15, 1945, and ends on August 9, 1945. Actually, there must have been some overlap back into the Operations period, but I have no dates.

6-15-45 to 8-9-45 = 2 months, rounded

Working Hours. Assume 7 day weeks, 14-hour days.

2. RATIONAL WARTIME.

I assume that the same overtime would have been worked, but that the activities would have been spaced out into a non-parallel, semi-sequential schedule.

2.1 Science. Science would have maintained the same schedule as the actual wartime schedule.

2.2 Conceptual/Preliminary Design. Design would not have started until the Feb. 16, 1943 reactor decision, and would have stretched until approximately three months after the June 1, 1943 bismuth phosphate separation decision.

2-16-43 to 9-1-43 = 6½ months

2.3 Semiworks. This would have begun at the end of Conceptual/Preliminary, but with a two-months overlap backwards.

7-1-43 to 7-1-44 = 1 year
7-1-44 to 3-1-45 = 8 months
Semiw'ks duration = 1 year 8 months

2.4 Final Design. Probably would have overlapped back into semiworks, maybe two months for site-work design. Note that plenty of design-affecting changes were happening in the last six months of the semiworks time frame. See Fig. 5.4

1-1-45 to 1-1-46 = 1 yr.
1-1-46 to 9-1-46 = 8 mos
Final Des. Dura. = 1 year 8 mos.

2.5 Procurement and Construction. Final site-work design would have been completed by 8-30-45 at the latest.

8-30-45 to 8-30-47 = 2 years
8-30-47 to 11-30-47 = 3 months
Proc. & Constr. Dur. = 2 years 3 months

SUPPORTING CALCULATION
FOR
REPORT ON MANAGEMENT OF THE HANFORD ENGINEER WORKS IN WW II

CALCULATION SUBJECT: DETERMINE PLUTONIUM-PROJECT SCHEDULES

DATE: July 23, 1994	CALCULATION NUMBER	3	Page 4 of 6

2.6 <u>Operations</u>. Would have begun 4 mos. before construction completion, as in the parallel schedule.

$$7\text{-}30\text{-}47 \text{ to } 3\text{-}30\text{-}48 = 8 \text{ months}$$

2.7 <u>Bomb Test & Drop</u>.

$$3\text{-}30\text{-}48 \text{ to } 5\text{-}30\text{-}48 = 2 \text{ months}$$

2.8 <u>American Invasion of Kyushu to Drop</u>.

11-1-45 to 11-1-47 = 2 years
11-1-47 to 5-30-48 = 7 months
Bomb delay   = 2 years 7 months

2.9 ^{238}U Bomb. In the Rational Wartime schedule, I assume that Oak Ridge would have followed the same rational method; thus the end-date relationship of the two bombs would have been the same as actual.

---

3. RATIONAL PEACETIME. I ran this calculation to see how it would turn out, but I don't really believe that there would have been an opportunity to carry it to completion without the need to convert to a parallel schedule. Seven months after Seaborg's demonstration of fissionability of ^{239}Pu, for example, Pearl Harbor happened!

In this schedule, all elements of the wartime schedule are lengthened by the ratio of the wartime overtime hours to a peacetime 40-hour week.

The only peacetime perturbation that I allow for is a presumed three months between most of the elements to allow for Congressional and AEC oversight. I made no allowance for labor troubles.

3.1 <u>Science</u>. $\dfrac{7 \times 10 \text{ hrs.}}{40 \text{ hrs.}} \times 42$ mos. = 74 months = 6 years 2 mos.

$$3\text{-}28\text{-}41 \text{ to } 5\text{-}28\text{-}47$$

I assumed that the Bill authorizing the scientific effort would also authorize conceptual and preliminary design in order to provide an accurate cost estimate for the next steps, prior to their authorization. I also assumed C & P design and Science ending on the same date.

3.2 <u>Conceptual/Preliminary Design</u>. $\dfrac{44}{40} \times 4$ months = 4.4 months
Use 5 months

$$12\text{-}28\text{-}46 \text{ to } 5\text{-}28\text{-}47$$

SUPPORTING CALCULATION
FOR
REPORT ON MANAGEMENT OF THE HANFORD ENGINEER WORKS IN WW II

CALCULATION SUBJECT: DETERMINE PLUTONIUM-PROJECT SCHEDULES

DATE: July 23, 1994	CALCULATION NUMBER 3	Page 5 of 6

3.3 **Semiworks.** Now we're spending capital $ and for that in peacetime we're talking Congressional and AEC oversight so for this element, Final Design, and Procurement & Construction the three month interval applies.

$$\frac{44}{40} \times 5 \text{ mos. design} = 6 \text{ months, rounded}$$

$$\frac{54 \times 2 \text{ shifts}}{40} \times 15 = 41 \text{ mos. for construct & operate}$$

$$\underline{+6 \text{ mos. design}}$$
$$47 \text{ mos.} = 3 \text{ years 1 month}$$

8-28-47 to 7-28-51

3.4 **Final Design.** $\frac{44}{40} \times 20$ mos. = 22 mos = 1yr. 10 mos.

10-28-51 to 8-28-53

3.5 **Procurement & Construction.**

Wilmington procurement went from 11-19-42 to 3-22-43, the start of construction. After that, Wilmington procurement went on in parallel with construction, but the time duration was governed by construction duration. This brings up the interesting point about cost stretchout that I go into at the end of the calculation.

$$\frac{44}{40} \times 4 \text{ mos.} = 5 \text{ mos. rounded. Before constr. start.}$$

$$\frac{54 \times 2}{40} \times 22 = 59 \text{ mos.} + 5 \text{ mos. proc.} = 64 \text{ mos.}$$

This is/interesting. When I asked Matthias how long he thought HEW construction would have taken in peacetime, he said, "five or six years". (127.FM)

11-28-53 to 3-28-59

3.6 **Operations.** They would work three shifts, as in the actual wartime case, so 8 months duration would be the same, until production of one test bomb and one for-use bomb. And, once the first reactor was done, they would have started using it, just as in wartime, so the same four and four relationship holds.

11-28-58 to 7-28-59

SUPPORTING CALCULATION
FOR
REPORT ON MANAGEMENT OF THE HANFORD ENGINEER WORKS IN WW II

CALCULATION SUBJECT: DETERMINE PLUTONIUM-PROJECT SCHEDULE
DATE: July 23, 1994   CALCULATION NUMBER   3   Page 6 of 6

3.7 <u>Bomb Fabrication & Test (No drop)</u>.

$$\frac{7 \times 14}{40} \times 5 \text{ wks} = 13 \text{ wks}$$

$$= 3 \text{ mos.}$$

7-28-59 to 10-28-59

3.8 <u>Project Stretchout Costs</u>. Extra time on a project costs money, as Barry and Paulson note, and as everyone observes. When in this calculation we go from 2.25 wartime construction and procurement years to 5.33 peacetime years, the procurement is stretched to match the governing construction time. You're buying the same equipment over twice the time. True, you do it with a reduced force, but there's certain overhead that multiplies. This is just one example, and there are many others. The conclusion is that the five-year job will cost a lot more than the 2-year job for building exactly the same thing.

<u>THE THREE SCHEDULES ARE SHOWN IN FIG. 5.5</u>

SUPPORTING CALCULATION
FOR
REPORT ON MANAGEMENT OF THE HANFORD ENGINEER WORKS IN WW II

CALCULATION SUBJECT:	DETERMINE NUMERICAL COMPARISONS AMONG MAGNITUDE DATA OF CHAPTER 2.
DATE: Oct. 3, 1995	CALCULATION NUMBER 4 Page 1 of 1

OBJECTIVE: Calculate various comparisons of Hanford magnitudes concerning plant area, excavation quantity, cooling-water quantity, peak employment, and intra-plant bus miles travelled.

1. Rhode Island Comparison.

   $$\frac{\text{Hanford } 670 \text{ mi}^2 \text{(Endnote 16)}}{\text{Rhode Island } 1058 \text{ mi}^2 \text{(High School Encyclopedia Golden Press, NY, 1961)}} = 0.63$$

2. Excavation Comparison.

   $$\frac{\text{Hanford } 25,000,000 \text{ cu yds (Endnote 29)}}{\text{Panama Canal } 262,000,000 \text{ (McCullough A p.617)}} = 0.095$$

3. Cooling/Domestic Water Comparison.

   30,000 gpm per pile; DuPont A, p.49

   100 gpcd was the domestic-water standard in 1943 prior to garbage grinders, dishwashers, daily showers, and large suburban lawns and gardens.

   $$\frac{30,000 \text{ gpm} \times 3 \text{ piles} \times 60 \text{ min} \times 24 \text{ hr}}{100 \text{ gpcd}} = 1.3 \text{ million residents}$$

4. Employment Comparison.

   Peak Hanford employment was 45,000 according to most sources, e.g. Carpenter A, p.4 I use this number throughout.

   Grand Coulee peak payroll was 5500, Oct. 1938, from records of Consolidated Builders Inc., Henry J. Kaiser, President; found in Box 500, Kaiser Engineers Record Center, Oakland, Calif.

   Panama Canal peak payroll was 47,500 per McCullough A, p.559

   $$\frac{45,000}{5500} = 8.18 \qquad \frac{45,000}{47,500} = 0.947$$

5. Round Trips - SF to New York City.

   2899 miles - SF to NYC scaled from Nat. Geogr. Map of U.S. (1:4,561,920, 1969)

   $$\frac{340,000,000 \text{ (Endnote 33)}}{2 \times 2899 \times 45,000} = 1.3$$

Appendix H

INTERVIEWS

INTRODUCTION

I interviewed these men in the order presented here. The Matthias interview is the longest because I knew little about the HEW at the time of his interview, and therefore required a lot of exploratory questions.

I asked a number of questions of these men apparently unrelated to management because I was either feeling my way or because I was seeking input for my section on technologies of the forties.

App. No.	INTERVIEW	Page No.
H-1	Franklin T. Matthias	159
H-2	Walter O. Simon	185
H-3	Raymond P. Genereaux	195
H-4	John B. Tepe	203
H-5	Russell C. Stanton	209

Appendix H-1

Interview with

COL. FRANKLIN T. MATTHIAS
AUS, RET

Col. Matthias was the Corps' Officer In Charge of the HEW. Prior to that assignment he was General Groves' Deputy Manager of construction for the Pentagon; subsequently managed major civilian heavy-construction projects; retired as Vice President, Heavy Construction, Kaiser Engineers. Immediately prior to his Hanford assignment he had participated in the two-week site selection trip in December of 1942 that culminated in identifying Hanford as the most suitable of more than 11 candidate sites.

The interview was conducted in three sessions at Col. Matthias' residence in Danville, California in the summer of 1993, and was supplemented by a number of telephone and lunch-time conversations in 1992 and 1993.

I used Matthias' Hanford diary to provide site-trip dates and times. The diary was also the basis for the parenthetical comments on interview/diary anomalies.

Plutonium-project veterans, other friends, and those familiar with the project were saddened to hear of his death at 85 on December 3, 1993.

Appendix H-1

TABLE OF CONTENTS

Chapter Outline No.	Topic	FM Comment No.
—	The Nov. 14, 1942 Wilmington meeting and the site-selection trip	1 -16
	Site report	13
	Site candidates list	14, 15
—	Interim activities	17 -21
—	Soils exploration	112
—	Construction activities; the camp	27, 44 -53, 108-111
—	Reactor startup and poisoning	56 - 59
—	Plutonium delivery	59 - 65
5	The Nature of DuPont	25, 28, 29, 42, 43, 89, 90, 101
5	Hanford Management	
	Superintendent/Project Mgr. relationship	28, 29
	Critical path	29, 38, 41, 42, 43
	Engineer management of crafts	30 - 37
	Oral communications; meetings; decisions	82 - 85, 94, 95, 106
	Decision making	106
	Management documents	89 - 93
	Design changes	102 - 104
	Lost time due to security clearance; priorities; expediting; equipment procurement.	115 - 117, 122, 126
	Schedule	66, 67, 69
	Governmental relations	96 -100, 107
	GAO at Hanford	93
	Single Authority	103, 127 - 130,

8	Labor Conditions	
	Productivity	75, 76, 124
	Disputes and labor relations	77 - 81
	Labor skills	118 - 120
	Turnover	54, 55, 123-125
	Wage scale	121
	Shifts and hours	70 - 74
	Peak employees	54
10	Intangibles	
	Conditions of mutual trust	24 - 26, 83, 84, 85
10	Additional Anecdotes	133 - 140

THE FRANKLIN MATTHIAS INTERVIEW

Harry Thayer (henceforth HT): You mentioned one day last year that in December, 1942 you made a site-investigation trip to the West Coast. Would you describe your early days on the project – what led up to the trip – and about the trip itself?

1. Franklin Matthias (henceforth FM): Now I think I want to tell you that Groves sent me to Wilmington on December 14 to see DuPont. He didn't tell me what it was about, but he said that the meeting was with some very high-grade people, and some high-level DuPont people, "And you go there and listen".

HT: And you didn't know what it was about at the time?

2. FM: I didn't know. I took the train to Wilmington. I didn't have the vaguest idea of what was going to be there. And he said, "Don't take much in the way of notes because this is a very great security problem; just try to remember all you can." And he said, "When you get through with the meeting, come back to Washington and no matter how late it is, call me at the office and I'll come down and pick you up". And he never forgot that. Any time we talked about that he'd talk about the strange thing of a Brigadier General going to get a Lieutenant Colonel!

At the meeting they had a whole bunch of scientists from the Chicago Metallurgical Lab that were there working out the health hazards from radiation. And they didn't know much because they didn't know very much of anything about the program, except that you could make plutonium.

That was discovered by a group of people at the University of California under Seaborg. He was a young chemist and he really developed the concept of the Hanford program.

So anyway we went through this session with all the hotshot scientists at Wilmington, and they would go through all kinds of things, and they would say, "Now understand, this is order-of magnitude", every time they made an estimate, "because we don't know". "But this is what we've got from our labs, and it gives us an idea, so we can't be very precise about this. " Which meant of course, that they didn't know to closer than a factor of ten, plus or minus.

That was quite a day! Now the Manhattan people that were there were Nichols and me, and that was really the first time that I got to know Nichols very well. And we had a whole bunch of hotshots, directors and everything from DuPont at that meeting, and we had Gil Church and Al Hall. Gil Church was called back to participate in this. He was then project manager of an ordinance plant, I think in Minnesota, that DuPont was designing, and Al Hall was their chief civil engineer in their design department. They were the three of us that made the trip to find the site. We didn't know that then, but the next morning they were sent right down to Washington, and the three of us went through with the Corps of Engineers and gleaned all the information they could give us. It was all kind of vague.

After the meeting I took the train to Washington and Groves met me at the station and we got in his car and we drove around Washington for about an hour, and the first thing he said was,

"What did you think of the meeting? " And I said, "Well General, I don't know what to think about it. I think I ought to get some Buck Rogers comics to get in the right frame of mind for all that science fiction."

HT: What was his reaction to that?

3. FM: He laughed – he thought that was great! Anyway, we kept on talking and he told me what the project was, and how serious, and this was a bomb. So he said that the next morning the two top DuPont guys would be in Washington to start out to find a place to build it.

Anyhow, the next day we spent with the Corps of Engineers – checking up on the water and power needs, and where they were – with the experts in the Corps office. And we got a lot better picture of what the scientists were looking at. And the three of us drew up a preliminary site plan showing six reactor locations.

The next day we left at 4:30 PM for Seattle via Chicago where we were grounded until Dec. 18th. We were grounded again in Cheyenne for five hours, so we didn't arrive in Seattle until 4:30 PM on the 18th. We talked to Col. Park and Col. Wild in the Seattle District office about electric power data, and at 10:00 PM on the 18th we took a military train to Spokane. We met Captain Hopkins of the Seattle District there at 8:00 AM on the 19th and he had a car for us to look around with. We spent all that day in the Carpenter Hotel studying air navigation maps of eastern Washington and Oregon that Captain Hopkins had brought us. We made some transparencies for map overlay based on the preliminary site plan. On the transparencies we plotted the reactors and separation buildings at the clearance distances that the scientists required, so we could get an idea by examining maps whether a site was large enough, and whether there were three to four mile minimum clearances to highways and railways and towns.

So the next day we started out and we drove right up into Idaho and all over northeastern Washington – Mansfield and Bridgeport and the Coulee-Dam area. We didn't go the other side of the mountains because one of the requirements they set up was 200 miles from the ocean. We didn't quite meet that on an airline distance requirement, but it was close enough. Anyhow, we found the Hanford site, with Hopkins describing it as one of the best.

HT: And you found that by driving?

4. FM: No, we had located it and a number of other possible places on the maps, and we checked all the ones up in the north. We spent the night of the 20th in Mason City, the construction camp that the Bureau of Reclamation ran at Coulee Dam. So then we drove some more around Washington because the first day we were there it was foggy and we couldn't see as much as we would have liked. Anyhow, we really covered eastern Washington pretty good – Soap Lake and Ephrata and Ellensburg – and we got to Yakima late afternoon on the 21st, and we spent the night there.

I was going to get an airplane from the Army Air Force base at Yakima, so the next day on the 22nd we went right over to the base because I was so sure that we could get one, and I had planned on all three of us making the trip to hit a few places down in northern Oregon which we thought were potential sites. And when we got to the airport, they wouldn't take 'em because they were civilians, but they could take me.

So they took a car at 10:30 AM and did some more checking in the Washington area, and then we were going to meet about the middle of the afternoon at the Navy air station at Pasco. And they were going to come in from the Yakima side through the Hanford site. So I made the flight down there, and they gave me a nice little plane that had a transparent floor that I could see down through – a photographic plane – and it was a great trip – the Deschutes River area and Madras – but I didn't find anything very impressive there, and I didn't go into the western side of Oregon either. But the places we had pegged on the map just didn't look good.

And I came back over Horse Heaven – in the area northeasterly from Plymouth – and over Rattlesnake mountain to the Hanford site from the west, and I got over that mountain, and I had looked at everything else, and I knew that was it, right then. We made a trip all around the river there so I could see the whole thing.

HT: So the Oregon sites never came to anything more than a map exercise – once you saw them from the plane, you said, well forget it?

5. FM: Yeah, that's about it. Those sites had too many people, or lousy topography, and power wasn't available in those areas – it would have been a long haul. Also they would have required a large pumping head and it would have cut into productive wheat land.

HT: And Horse Heaven would have required more pumping power for river water.

6. FM: That's right, and the terrain would have required much more earthwork.

And then I met the DuPont guys; they had come into Pasco through Yakima from the west – there's a west highway through there – it's not a very big one, a pretty good one – oiled – and when I met them at the airport, they were just as excited about that Hanford site as I was. And we were sure that was it!

We were convinced that we had found the only site in the country that was that good, and I still think so – no other place has all the good features of Hanford. Rattlesnake Mountain for one thing, provided a barricade from the rest of the state, and so few people – just those along the river.

HT: It was obviously a good one. What was wrong with the site near Coeur d'Alene?

7. FM: Well, too many people, and the construction topography didn't seem to be very good in the areas that didn't have too many people, and it was quite a far chunk for the transmission lines. It just didn't have the kind of things that we found at Hanford.

HT: And the one northwest of Coulee Dam you once told me would have required a lot of pumping power.

8. FM: It would have caused them to build the big pumping station at Grand Coulee. Because we had to lift water about 2000 feet. This was north and west of Coulee.

HT: And they never completed that pump station until much later – after the war, I think.

9. FM: Oh no, and this would have been a tremendous cost. And at Hanford, having that Columbia river down in front of you was a very important piece.

HT: A hundred thousand CFS, and pure; snowmelt.

10. FM: Even in drought it's about forty thousand.

When we left Pasco I called General Groves and told him I thought we'd found the place, and I think we're all done – we're coming home. But he said, "Have you got some more candidates?" And I said yes down in California, for one thing, and in southern Oregon there's something that we thought might have possibilities. So he said, "You have to go down to the other sites; otherwise someone in the future can say we went off half cocked. You might miss something down there, or we might have difficulty getting one of the other sites, so you better go down and look at them".

So we left Pasco at three AM on the 23rd and got to Portland at 11:10 AM and went directly to the Portland District office. We met Col. Leahy and Lt. Col. Tudor and discussed the Deschutes and John Day areas as being the most suitable in the Portland District. We left on United Airlines the next day for Sacramento, but the flight was three hours late so we were too late for work at the District office, so at 6:30 PM we drove to San Francisco. We laid up there and reviewed everything we had, and what we still had to do, and had Christmas dinner at a Chinese restaurant in San Francisco. I called Col. Hunter in Sacramento and asked him to look into possible sites in his District, and gave him the criteria and arranged to talk to him the next day.

On the 26th we talked to Col. Hunter in Sacramento about a possible location on the Pit River. We tried for a recon flight from Mather Field, but fog and rain in the Pit River area prevented us from going there. We considered driving but the weather forecast was for cold and snow; it was sleeting at Alturas at the time. Hunter promised to write us a report on the area and mail it back to us, but later on we thought it over and told Hunter that he needn't be too serious about that. We had a second "Christmas" dinner with Col. Hunter - then took the night train to Los Angeles, arriving there two hours late at ten AM on Sunday the 27th. We spent that day reviewing our notes.

We talked to the Los Angeles District Engineer on the 28th about power availability in the Blythe area, and found there was little to spare. We made a phone call to Giroux in the Chief's office in Washington, D.C. and he confirmed that power there was very tight. (Here, there's a difficulty about the diary entry. According to the diary, the LA District reported power available, but that Giroux did not phone back before the site group had to leave for Blythe. Later in the interview Matthias stated that the Blythe site was rejected for a security reason, so power availability, or the lack of it, obviously became moot anyway.)

We tried to arrange an overflight of the Blythe area, but bad weather prevented it, so we drove over to Blythe, arriving at 10:30 that night.

They put us up in an Air Force BOQ near Blythe. Al Hall was a good deal older, and he was cold, so I put my overcoat over him and he went to sleep. About midnight some of these guys came in and claimed we were in their bunks. And the first one they went to was Al Hall, with my overcoat and insignia over him. And they started to dump the bed to get him out of there. And they grabbed my coat with the insignia on it, and boy, they hightailed out of there!

HT: That's funny.

11. FM: Yeah, they weren't going to fool around with a Lieutenant Colonel! Next day, we got a five-place Cessna and flew the area south of Blythe. It was all irrigated and developed.

We also drove the Blythe area and the desert, and for a building site it wasn't bad, with Colorado River water, but, as I pointed out to General Groves in our report, if we wanted to keep this secret, we didn't take any water from the canal.

HT: The farmers were sensitive.

12. FM: Well sure, we'd never get by with it – keeping it quiet. There would be so much hell raised. And anyway, we didn't have that problem on the Columbia. (The diary records that the site team also flew over an area near Needles and that topography appeared suitable, but says nothing about why Needles was rejected. It is possible that the same water-withdrawal problem associated with Blythe mitigated also against Needles.)

From Blythe, I called the LA District about hotel rooms. We drove over to LA and spent the night – the 29th – in the Biltmore.

I should say here that Groves had made arrangements with all the District offices. He told them I'd be asking for information, and not to be too inquisitive about why they want what they want, but to help them. So we had some good support and transportation and everything.

Anyway we left LA on a 2:50 PM flight on the 30th and arrived in Washington, D.C. at 10:30 AM on the 31st.

HT: Did you write a site-investigation report?

13. FM: The three of us did a handwritten report on the plane going back, but I'm not sure how much of it ever got typed.

HT: So, just to recap, the most important sites you had marked on your maps during the Spokane meeting were: Coeur d'Alene; NW of Coulee; Mansfield; Hanford; Deschutes River-Madras area; John Day River area; Horse Heaven; Pit River; and Blythe. Were there any others?

14. FM: There were, but I don't remember them now.

(The diary mentions Moses Lake as a possible site but says nothing more about that site, for or against. It was sure to have been rejected for the reason of adverse pumping power, if for no other.)

HT: You mentioned Soap Lake, Ephrata, and Ellensburg. Were those just places you drove through?

15. FM: That's right, they weren't sites.

HT: And on January 1, 1943 you met with Groves to report on your trip?

16. FM: Yes. Besides General Groves in the meeting, was Major General Robbins – the Deputy Chief of Engineers, and Carl Giroux They were in general agreement that the Hanford site was it. Giroux thought that Hanford was perfect for power.

General Groves reserved his final approval until he saw the site. He thought he had to see it – to make that monumental decision. He planned to go out on January 19th, or thereabouts.

I also reviewed the whole thing with DuPont in Wilmington and with Nichols in New York, and I returned from that meeting on the 10th.

On the 11th, Gil Church and I and DuPont's Grady – their proposed operations manager – and two real estate officers from the Chief's office flew to Spokane to meet a real-estate specialist from the Portland District office. We then drove to Hanford so the real-estate people could look it over.

HT: And to show the real-estate people the land requirements you used your preliminary site plan?

17. FM: Yes, and we used that preliminary plan for quite a long time.

HT: When did you get to Pasco for good?

18. FM: Not right away. The next thing was I left for Pasco to meet Groves there. I left for Pasco via Chicago and Minneapolis, but on January 15th the plane was grounded in Bismarck. So we took the train the rest of the way and got off in Pasco at 4:30 AM on the 17th. I missed seeing Groves because he left that same day on the Northwest Limited for Chicago.

On the 18th I looked over the Prosser area and Benton County. On the 19th I drove to Pendleton and caught a plane to Washington, and I got there on the 20th.

But I had arranged for a real-estate man from the Portland District to meet Groves at Pasco, and I told him about the land requirement we had, and that Groves would want to know just how to proceed with all that, and so it was alright.

HT: What did you do back in Washington?

19. FM: Right after that we got military approval for the site, and we did all kinds of things in getting some design going. It wasn't until February that General Groves called me in. I'd been doing a lot of things like getting aerial flights cancelled over the area, and getting the Army and Navy off – one of them was doing gunnery practice and the other bombing practice up in the Saddle Mountain area, and I had to get them off, and Groves had me do all this; I didn't know what my function was to be. Many, many months later he said he intended me to be the executive officer.

HT: You'd help him out in Washington, D.C.

20. FM: Yeah. Anyway, that's what I was doing. And I don't know exactly when, and I don't have any records, but I think it must have been about the middle week in February, and I'd been doing all the things it took to get started, but I had no authority and I had no assignment.

And Groves called me in and told me that the chief of Engineers, General Reybold, had told him that he could have any officer in the Corps who was not on some other real important duty to run the Hanford project. Groves told me that; then he said,

"Now I wish you'd do a little scouting around and find somebody we can get to run the program at Hanford."

And I said, "Alright, General, I'll do some checking up", and I started out the door, and I had my hand on the door – I can remember this so vividly – I had my hand on the doorknob, and he said, "Hey, by the way, if you can't find anybody you like, you're gonna have to do it yourself". And I opened the door wider again and stepped back and said, "General, there isn't anybody I can recommend". I wanted to get out of Washington, D.C.

HT: Did he see the funny part about that?

21. FM: Oh, sure he did. He said, "Alright, you're it – you're appointed right now. You're boss of the Hanford project".

So then we did a lot more specific things. We had some agreements to work out with DuPont. There were just a lot of things. We worked out what my authority would be with them, and what they could depend on, and what they could do on their own, and all this stuff. We spent probably a week on that – really working out the whole program – how we'd operate at Hanford. And they had appointed Gil Church as the construction manager for DuPont. And we were both about the same age. (34). We got along so well, right from the start, and we'd gone on this trip. And as a matter of fact, Groves many years later in his book mentioned the fact that he'd gotten Gil Church and I to go on that trip, because he wanted to know how they were going to get along if they stayed there. So he had that in mind before we made the trip.

HT: Is Gil Church alive?

22. FM: I think so, but I don't think he's all that well. I haven't heard anything in several years. Gil Church was great!

HT: Getting back to the agreements, did you have any Corps lawyers working on it with you?

23. FM: No.

HT: How about DuPont lawyers back in Wilmington?

24. FM: I don't know, but I doubt it. (But see parenthetical note following 26 FM.) This whole thing was put together with such faith people had in others – the leaders – it was just amazing.

Did I ever tell you when Groves proposed this project – that DuPont take it over?

HT: It was Groves' idea that DuPont take it over?

25. FM: Of course, because he had been in charge of construction of all the munitions plants. And DuPont was building a lot of them. They had done so much good work for him – amazing work for him!

Alright, he went to Wilmington and talked to all the hot shots. And he finally had to tell them what it was all about. And he went to a meeting in Wilmington the day they accepted the project, after he made his pitch. There were about 10 or 12 people at the table – the top of DuPont and Groves. And he had a folder with a lot of information on it at each chair. He opened the meeting by saying, "Now what's in front of you is a discussion of what this project is for. It's very, very critical information that you people should know, and we've talked about some things that are beyond our normal security limits. This goes into a lot of detail, and I'd rather you didn't read it, but you're at liberty if you have any doubts or questions at all, you can read it.

And there wasn't a single DuPont person in that whole group that touched that thing until they promised Groves they'd take it.

HT: I was talking to some Kaiser people about this problem, and one of the main things that engineers and contractors are worried about now, and spend huge amounts of time on, is to avoid getting sued – how do we write this up so we don't get sued – and nobody was worried about that at that time?

26. FM: Never. Not a bit.

(The great confidence in others certainly existed then, as stated by Matthias. Such confidence however, existed within strict limits. Before DuPont committed themselves in writing – the Nov. 6, 1943 contract for example, they made a thorough legal analysis, as noted in W.S. Carpenter's Oct. 21, 1943 remarks to his Board of Directors, and as supported by a three-page summary by DuPont's attorneys of seven risks for which DuPont demanded protection in their contract. The remarks and summary are on file at the Hagley.)

HT: So after you made these arrangements in Washington with DuPont, and so forth, then you went out to Pasco for good?

27. FM: I went out about the middle of March; that was about the time that DuPont had just signed their construction superintendent.

HT: Who was he?

28. FM: I can't remember his name, but he ran the crafts. He didn't have the authority that most superintendents have because they had all the engineers doing the stuff that the superintendents would normally do.

The superintendent was a real nice guy, and he did a nice job. He managed the construction stiffs and the engineers managed the job.

HT: A very interesting way of doing it – I've never seen it done that way.

29. FM: No, I haven't either. DuPont had a definite limitation – the superintendent was not the real boss of the project – Gil Church, the Project Manager was, and Gil Church had a staff of inspectors and planners, and all kinds of things.

DuPont at that time really had a program on construction, very thorough, and they cut no corners on any job.

I kidded them one time that if they had an outhouse to build they would use the same system as they did on their large jobs, and they thought that was very funny.

I had learned about how they'd do that because Groves at one time had me review some of the construction jobs they were doing, long before he was even with the Manhattan District. And he got me to review some of their ways of making reports, because they were so far ahead of any of the other builders in the country – he just wanted to know how they did it. So I had gone through that kind of a drill on my own – that's one of the personal assignments Groves gave me.

They – for the first time – they didn't talk about it then, but they were using the program of critical path.

HT: They started it that early?

30. FM: They started it but they didn't know what they were starting. But for each of the major projects – reactors, and the other big buildings – had an engineering staff with an engineer who was responsible to lay out every day, every job that every craft group had to do.

HT: Was this for the craft superintendent, or did this go down to the foreman level?

31. FM: The instructions went down to the foreman level, or assistant superintendent.

HT: Well, that engineer must have been pretty skilled in understanding construction productivity.

32. FM: All of that. The engineers would lay out every task for a day ahead. Then they'd supervise it. And the workers and the foremen were told just what to do by the engineer.

HT: So that's how it happened. How many people did DuPont have in that planning group that layed out the work.

33. FM: Oh, about four or five in each of the major projects, like each of the reactors, or each of the separation plants.

HT: Did DuPont's planning engineers do detailed written instructions for the water pumping stations – Atkinson's work?

34. FM: No.

HT: How about the power plants and the high-lift pump stations?

35. FM: The DuPont scheduling engineers wrote work instructions for those.

HT: Do you remember the details of this type of planning?

36. FM: I've always thought that the planning engineers came on an early-morning shift or an evening shift to write the instructions for the next shift. They didn't spend all day watching all of them. They did a lot of planning for tomorrow today. These engineers had to check after every shift.

I don't know the exact details of how these engineers did it but I know that they did it. Every foreman got his instructions as to exactly what he was to do that shift. It didn't always work exactly, and they had to make some adjustments sometimes. And they sure didn't have a bunch of pipefitters tearing down what the shift before them did because they didn't like it.

I think that as far as the overall efficiency, it was all because the crafts were guided in what they did and told what they had to do, not "you gotta get this done", but "this is what you have to do", and what part of it, and everything laid out for them every morning – every shift start. And they were watched every minute by the engineering group.

HT: Those were sure precise instructions.

37. FM: That was it, and I think that was the difference.

HT: No project I ever saw after the war had that kind of direction.

38. FM: No. Well, they never have had until this thing. Now, I think this led up to critical path scheduling and everything that's gone with it. This critical path program was announced first in the ENR, or Civil Engineering, I'm not sure which, written by Frank Mackie, who was one of the key DuPont people on this project. He was out of Wilmington and out at Hanford for a while – a very competent guy. He worked that up in 1960 – the real critical path philosophy.

HT: And it was based on his experience at Hanford.

39. FM: Absolutely. Now this was 16 years later, because I knew it in the early '60s when I was on a job for the Aluminum Company in Canada, and I spent about the last five months of the job – I took his report and I transferred our whole scheduling on a big hydro project, with a five-mile tunnel and an underground powerhouse – and I translated that – I did it myself – from his instructions, his descriptions for the whole program – the critical path and everything – instead of bar charts - and they were all related - everything was sequenced to something else, and that's what the bar chart never did.

HT: Did you also keep bar charts on the Hanford project?

40. FM: Yes, he kept bar charts, but he kept them from the basis of a critical path, so that he didn't miss anything that would hold up something else.

HT: And they actually laid out drawing sheets with activity lines intersecting at nodes, and parallel sequences and determined which sequences were critical?

41. FM: They sure did.

HT: Their organization was way ahead of its time – like 20 years. And from what you said earlier, that's because DuPont had begun to work out this method before they ever got to Hanford.

42. FM: They had it worked out. It's hard to peg these things, and why, but I've always thought that system of DuPont's was what let us do a very efficient job at Hanford.

HT: I'm convinced. I had no idea that they planned it that way.

43. FM: It was really new to me too. But I had done some work for Groves on the Pentagon, applying DuPont's system for reporting that was a cost-plus job, and I got them to process our monthly report, with all the subcontractors, and everything else in it under that same system, so I learned something about that before I went to Hanford.

HT: Would it be correct to say that before you started to build the process things, you did the infrastructure – the roads and railroads and the camp?

44. FM: That was the first thing – the camp – that we had to do. Our construction camp was the third largest city in Washington.

HT: Did you have your offices right there at Hanford, or were they down in Richland?

45. FM: During construction, we had our offices at Hanford, and so did DuPont. Before January, 1945 we moved our main offices down to Richland. We were vacating the camp then.

The roads were not much because we had a main road. We had to widen it, but we had good access right up to Hanford.

HT: You had to build a big railroad network in the plant?

46. FM: No, we didn't. The Milwaukee Railroad – that was one of the pluses of our site study – had a branch that took off from the east, north of Priest Rapids and crossed westwards over the river on a bridge to the right bank – north of Saddle Mountain – and came around the bend, right down where Priest Rapids dam is now. That was a fifty-mile spur line. It was in working condition, and they'd come in about twice a week and pick up freight and deliver freight, but we had to do a lot of rework to make it in better condition. We then extended it down to Richland. That's one of the things I had quite a discussion with General Groves about – I said we needed to do that.

HT: Was the substation near Priest Rapids there at the time?

47. FM: Yes. You know, when I first laid out what we were going to acquire, I included the Priest Rapids dam site – I knew it was being planned – because I thought it would be nice for the Corps of Engineers to control it, right then. But one of the executives of one of the power companies – I forget which one – found out it was in the Hanford Works and he kicked about it, because he said it was a power site. And we backed down – we didn't try to argue about it.

HT: Did you start the river-water pumping stations early?

48. FM: Well, our first work was the construction camp – we knew we were going to have to have a hell of a lot of people – so we organized and built quite an effort right away.

HT: When did that start?

49. FM: The middle of March. We didn't do much improvement of roads for several months - that was not a high-pressure thing because we had pretty good service as it was.

HT: How about your water supply – did that start about then?

50. FM: Oh no, we didn't start that for maybe a couple of months. But when we did start it was all under subcontract. Guy Atkinson did quite a bit of building of the intakes at the river and the main-line feeders to the different places.

HT: What was the first reactor you started?

51. FM: B. We had listed six reactors in our preliminary layout, A, B, C, D, E, and F. Then, when we cut it down, we just took advantage of a greater distance between reactors.

We started general site excavation for the reactors before we had dimensioned drawings. Wilmington would fill in the dimensions by teletype.

HT: When did that excavation start?

52. FM: We started it, probably, early in April. And I think the design didn't catch up to us until early June.

HT: Why did you start with B instead of A?

53. FM: The topography at A was not a good kind to build in. The river channel was kind of rugged and tight, and that's where the Indians lived.

HT: I saw a picture of the camp in one of the books, and it was a monster.

54. FM: Well, we had 45,000 peak employees in the middle of 1944 and we had hired 132,000 that went through the works. Quite a lot of turnover, but it wasn't bad.

HT: When I got to Hanford eight years later, they were still talking about a thing they called "termination winds".

55. FM: Oh yeah. We had as many as 500 guys leave after a dust storm. But that was not as big a thing as they've said – it's been exaggerated.

HT: Now on September 26, 1944 the B reactor started up. And then you had to shut down because of poisoning. When did you start into production?

56. FM: I think it was about six weeks that we got that fixed.

HT: The last blank in my chronology – when did you fly the plutonium down to Los Angeles?

57. FM: Alright, I've got some notes here. After the B Reactor failed, it took Fermi and Wheeler just a couple of days to decide what the problem was – real good detective work.

There had been a big argument between DuPont and the Met Lab. DuPont had designed the reactor for – I don't remember now the number – 2000 pipes. But the scientists said that was way too many – you ought to cut that down – it would save a lot of work and a lot of money. And DuPont said that this is a thing that we don't know for sure, and you don't know for sure, and we're going to build a little bit extra in the plans. And they made their point, and they added 500 pipes over what Chicago wanted. And that was just what it took to bring it up.

HT: They filled all 500 with uranium?

58. FM: Right. And it worked just exactly like they'd planned it that way.

HT: If it hadn't been for Dupont's engineers, we'd have been a couple of years behind schedule?

59. FM: Right. It took about six weeks to get the extra pipes filled, and of course all the others were filled with the same amount.

Now when I went to take the first plutonium to Los Alamos, it was the fifth of February, 1945.

HT: Anyway, you went down by plane on February 5th.

60. FM: I didn't go by plane. I drove from Hanford to Portland. I had a guy with me and we had a locked space on the train from Portland to Los Angeles.

The guy I had with me was my labor specialist; he was one of the military detachment assigned to Hanford – a great help to me – one of our staff of security guards.

HT: Can you tell me how big a volume of container you took from Hanford?

61. FM: Well, I would consider it about a two-foot cube, wrapped up in wrapping paper and ropes, and inside was a test-tube thing suspended and secured – all surrounded by lead and rigged so it stayed right in the middle of that box. It was quite a heavy thing, and I carried it just like a box any traveler might have with him.

HT: So it didn't look important.

62. FM: It didn't look important at all.

HT: So from about December of 1944 to February 5th of 1945 was how long it took to produce the first amount.

63. FM: Well, you know one of the things that hasn't been publicized much – the reaction of converting uranium to plutonium doesn't happen suddenly, and it goes through the reactor where it changes to neptunium, a step between uranium and plutonium. Then neptunium starts losing neutrons and changing to plutonium.

Now this is not a fast thing - it took about three months to get the optimum amount of plutonium. So we stored it in a pool about 20 feet deep at each of the reactors about three months. Well, early in 1945 we had taken them out of the pool earlier than that because Los Alamos was begging for plutonium because all they had was tiny lab samples and they wanted something more to prove that they were right. The only time I had telephone calls from Los Alamos was about then: "When are you going to get us some plutonium?" So we shortened the seasoning time, which is not the most efficient way, and processed the stuff so we could get some measurable quantities of plutonium down to Los Alamos.

HT: Did you make a second trip to take the bomb materials?

64. FM: No, the next trip was a little bit more, but the same way. I'd gone by driving to Portland and getting on the train and turning the package over to an officer from Los Alamos. I remember when I got there this guy was there to meet us. I said,

"Well, how are you fixed for space going back to Los Alamos?" He said, "I had a heck of a time getting a room, so I have an upper bunk". I said, "You know you can't do that – you have to get yourself a room so you can lock yourself in. You're going to have to understand that this stuff you're taking back to Los Alamos costs $350 million!" So he went back to the station and negotiated for a locked room. The next batch went down there about a week later – maybe two weeks later, and Major Riley, who was my administrative officer, took that by the same route, to Portland and then by train to Los Angeles, and then turned it over to a guy from Los Alamos.

After that we worked out a system for getting this stuff delivered, as we went into production on a larger scale. We set up a secure place at Fort Douglas in Salt Lake City, and Los Alamos had some guys quartered there, and we had some, and the two groups were isolated from each other. We would haul a load up in an Army ambulance from Hanford, with maybe a week or two's production. And we had the ambulance because we didn't want to attract attention, and there were Army ambulances all over. And they would drive down to Fort Douglas and deliver it. And we had a couple of security officers running that place. They would then take the load to another part of that reserved space and there the Los Alamos Army guys would pick it up and drive it to Los Alamos. The Los Alamos and Hanford guys never saw each other and the Hanford guys didn't know where the plutonium was going.

HT: Well, now I'm getting it clear. The story I read in one of the books that you got on a plane and flew to Los Alamos was not correct.

65. FM: No.

HT: Did you have an originally scheduled completion date when you started the job?

66. FM: Yeah, the contract was for four years, to go into operation. But we had a tighter schedule than that. We were just about operating on schedule. And we finished construction and went into operation – everything – sometime like March, 1945.

HT: You beat it by a mile. Four years would have taken you into 1947.

67. FM: And we had everything. Our three reactors were operating in January. DuPont got an Army-Navy "E" award in October of 1945 for their work on the project. I have a tape of the celebration when they were given the award.

The president of DuPont was there and he was telling us his impressions of the job and all that, and he said, "You know, we finished the job and turned it over as an operating facility, and the Government reduced our fee from $1 to $0.68, the reason being that we were a year early in contract completion". And he said, "A few months ago I got a letter from the Pasco Kiwanis Club and this letter had in it $0.32. The letter said, 'We heard that you've been cut in your fee, and we think that you did a pretty good job, so we're going to round it out'. And they enclosed a check for $0.32, and that's posted in my office".

HT: Were the auditors being funny?

68. FM: I don't know – I think that was part of the contract.

(Groves, when he wrote of this incident, said the Federal auditors were serious!)

HT: Obviously in your planning you had intermediate completion dates.

69. FM: Yes, and we kept working and realizing them, as things developed. And we gained on many of the intermediate dates. We were a little disappointed in the speed with which we could translate the exposed uranium that was kicked out of the reactor – the time it would have to rest to get an efficient conversion to plutonium. That slowed us down, and we did everything to try and beat it, and we processed quite a lot before it was up to optimum.

HT: Getting over to your shifts – you had two shifts, I think you said once.

70. FM: We had ordinarily two shifts. And we'd work extra shifts sometimes when the job was falling behind.

HT: And you worked six-day weeks?

71. FM: Yeah, and sometimes we'd work Sundays.

HT: And the shifts were how long?

72. FM: Eight hours to start with – then we went to nine, and they got paid for time to and from the job.

HT: Were your shifts head to tail, or did you have an hour or so in between shifts?

73. FM: I think we had an hour between shifts.

HT: And you and Dupont's management and engineering worked nominal eight or nine hour shifts?

74. FM: Yes. Most of us worked more than that. And Dupont's engineer-controllers worked any time the crafts were on the job.

HT: The craft labor – what percentage of the nine-hour shift would you say they actually worked?

75. FM: Well, I wouldn't know how to say, but I think the fact that all those engineers were running around checking up – I don't think they spent much time loafing.

But I have to tell you a little story. Most of these people lived at the construction camp. We had buses that picked them up in the morning and then brought back them back, and one time we couldn't get nice buses. They weren't available in the United States. And we had a bunch of standup buses – that's all we could get. And we put 'em to work, and the crafts didn't like the idea. So the DuPont transportation guy figured out this system. We'll take these bad buses and put 'em in line at the load-out lines for the morning shift at the rear of the good buses. And then coming home, at the loadout lines at the working area, the good ones were last. So any one quitting early had to stand.

And another thing; at one point we didn't think we were getting as much productivity as we should have. Now, we got an improvement of that by sort of a strange situation. We had a terrible time getting pipefitters right at the peak of the project. And the pipefitting was a tremendous job because all of those pipes in the reactor had to be rigged to bring pressure readings from both ends back to the control panel. So we had a tremendous amount of pipefitting, and we were desperate in the middle of 1944 for getting fitters. We went all over, and the unions tried their best.

Jack Madigan was a retired, dollar-a-year consulting engineer sent out by the War Department to look at Hanford, Oak Ridge, and Los Alamos. I told him about this terrible pipefitter shortage and discussed some ways to solve it, and he made this suggestion. "Look, we've got a lot of pipefitters in the Army on limited service kept in this country in the Reserve Army, and I'll bet we could get a couple of hundred of those guys out of what they're doing, and out to Hanford, and I'll check up on how we could do it", and he called me up the next day and said, "Well, it's simple. We've got 'em all in our computer." They had a computer in the Pentagon. And he said, "We're going to find out where we can get the pipefitters. I've already talked to the labor union and they will approve any qualified people, whether they're members or not, to come out and try to help to catch up."

So it only took about three days and they were sending guys out from army stations all over the darn country, to Walla Walla at the Army hospital there to put them on active duty.

I met the first batch of about thirty that were going through processing there, and I told them, "You've got an opportunity to work on a project that's extremely important, and desperately needed. We've got grants from the unions, and you don't have to fool with union problems." A lot of them weren't union men at all. And they were all skilled workmen. I told them, "Now you're going to work here and you're going to get regular pay, and you're going to go off active duty, and if you do what you're supposed to do, you're going to stay here until we're through, which may be a couple of months or more, and then you can either be free from the army or you can go back on active duty. And if you don't perform well, either individually or collectively, all I have to do is phone one telephone number and you're back on active duty."

HT: Sort of an incentive plan.

76. FM: Yeah, it was kind of an incentive plan. Now I said to them, "What I hope you can do is to make the pipefitters on the job feel that this is so important that they really better get to work and maybe we can get some more incentive into the program." So, in about a week we had about 200 of them. And we put them to work. I worked it out with the DuPont company. We didn't put them in separate squads, but mixed up with the regular pipefitters so they could get into the thing, so they could tell them what kind of a deal we've got and how great it is and how important this must be, so that maybe we'll get some better productivity, and we did. Instead of six weeks catch-up, we caught up in a little over three weeks.

And when we were finished with them, and got up to schedule, we sent a bunch of them down to Oak Ridge, and they did the same thing for Oak Ridge.

Jack Madigan went back to Washington and reported to the Undersecretary of War about his trip and said, "You know, this is the best thing that ever happened to us. If it works, we won't have to explain anything about it. If it doesn't work, we won't have to explain about anything else!"

HT: How much time was lost in labor disputes?

77. FM: Very little. The pipefitters once sat out for an hour and a half one Monday morning. I'd taken that weekend to go fishing. The guy that built a lot of the housing was going fishing at Astoria at the mouth of the Columbia River, and Riley – my administrative assistant – and I drove out there Saturday night – we got there about midnight to go fishing at five o'clock the next morning. About 10 o'clock a police launch came out looking for me – I had an important call from Hanford. So I went to the phone and the pipefitters were going to go on strike Monday morning. And so we drove back right away that night. It was quite a haul – driving – and we got there about midnight. The meeting was next morning.

And I took the stage in the recreation building – it held a thousand people – and they had loudspeakers to talk to themselves. Apparently they had a date with a union guy – an official – to be there to help them. There was one guy on the stage that I knew – a young fellow that was a pipefitter, and I said, "Give me a chance at that microphone". And he said, "Good, these guys sure need something."

So I gave them a little talk, and I said, "You know, you guys are breaking your promise. Your union agreed that we'd have no strikes. Now far more important than that, you are having a strike and not working on a project that is of tremendous importance to this country. Now, I thought you appreciated that, but apparently there some of you that are moving this strike, and they're working for the Germans, and if I knew who it was that was fomenting this thing, I'd arrange to have them sent back to Germany, where they belong."

HT: How did they take that?

78. FM: If they had guns they would have shot me! And after I got them quieted down I said, "Look, take it easy. I'm not calling you traitors, but some of you are acting like it. Now how about going back to work and doing what you promised, and what we need badly. I'll have the buses at the door in ten minutes." And then they cheered. And that was the end of it. They all went back to work, and their union leaders came as they were rushing out the doors, and they wouldn't go back in to listen to them. And I said, "I'll set up a deal, and we'll discuss all these problems you have and we'll do something about it, and you set up a committee, but I don't want a committee that comes from guys that are hardly working. We need to be working."

The DuPont guys thought, "Gee, that's a dangerous thing for him to be doing."

HT: Yeah, well in normal circumstances it would be.

79. FM: Ordinarily, their union agreement was with DuPont, but we had this kind of relationship. I knew I could do it more than they could.

We had another electricians threat later, but it never amounted to a strike. We settled our discussions with the pipefitters the next afternoon. So I can't give you even a reasonable estimate of time lost; there was practically none.

HT: Well, that's probably because of the time you spent on labor relations. I've been reading one of the public copies of your diary, and it appears that you spent about one quarter to one third of your time talking to business agents or international vice presidents. So that time seems to have paid off in the easy labor relations on the job.

80. FM: Well, I did talk to them a lot, and my door was always open to the business agents and stewards.

HT: I was surprised at the frequency of your contacts with Dave Beck.

(Dave Beck was the teamsters leader in Seattle, and very well known nationally because of the fascination of the newspapers with this very charismatic man.)

81. FM: Dave Beck was a very cooperative guy. Whenever we decided in a meeting that he and I were going to do things in a certain way, then he always did the very best he could to do what he promised.

He ran this little newsletter for the teamsters under his jurisdiction, and I'd read it and once in a while I'd find something in it that was not right and I'd phone him and tell him, and he'd say, "Well you write it the way you think it is". And I would and I'd mail it to him, and he would always print it in the next newsletter.

HT: When you communicated with your staff, was it mostly oral?

82: FM: Yes. Almost 100%. I remember once in a while writing them something.

HT: How about talking to DuPont? Was that oral?

83. FM: Very little writing. We had an unusual respect going, and Gil Church was great.

HT: Did you write confirming memos to them?

84. FM: No. Everybody trusted each other on the management side.

HT: Did DuPont write memos to each other?

85. FM: Very rarely would they write. They'd have frequent meetings – informal usually. In my dealings with them we didn't have to write a bunch to protect ourselves.

HT: Well, that's one reason you got the thing done so fast.

86. FM: Sure it is. You know, Slim Read – that fat guy up there in that photo (here Frank pointed to one of the many photos on the wall of his study) – he was boss of DuPont's construction, nationwide, and he was talking to General Groves one day and he mentioned the fact that I was having quite a tough argument with Gil Church, and Groves said, "That's great – if they didn't have arguments sometimes, I'd fire them both". We had a really wonderful relationship with the DuPont people.

The guy that was sort of handling things for Gil Church in Wilmington was Frank Mackie. And he was the one who really wrote up the critical path method. And Gil complained that he had so many things that Mackie had to approve in Wilmington, that I promoted the idea that Mackie be moved out here. And then he had to go to Wilmington to get authority! And then I kidded him that authority in DuPont was inversely proportional to the distance from Wilmington!

But this was all under what they called their Explosives Department run by Roger Williams. I had a lot of contact with him and I had to bug him about not getting something done in a hurry. Did I ever tell you when I left the Army?

HT: No.

87. FM: I stopped in to see a lot of the DuPont people when I was on my way down to Brazil. And I talked to Slim Read and he and Gil Church and a whole bunch of other DuPont people – I had a nice visit in Wilmington; spent all day and had lunch with them. And at that time they were just starting in the Savannah River project, and Slim Reed said to me, "You know, the things we have to put up with on that project! We never knew how good we had it at Hanford!"

HT: I'll bet. Already the AEC had started clamping down.

88. FM: Everything was crowding in – all the government controls. Anyhow, I dropped in to see Roger, and we had a nice visit, and I think at one time I think the thing that he was annoyed with me most was that we were behind in getting some important information from them for the construction, and I sent him a teletype and I asked him to see what he could do about getting that done, in a hurry because we needed that badly. And I said, "It is being noised around Hanford that it takes DuPont women ten months to have a baby". And that kind of burned him up! But he was a nice old guy.

But when I left, here's what he said, "You know, we've done a lot of awful tough things together, and I hope you feel as I do, that our differences were of the mind and not of the heart".

There are not many relationships like that now in the business.

HT: That explains a lot. Getting into your management documents, did you have a Chart of Accounts?

89. FM: Yes. The DuPont accounting system had a standard breakdown of construction accounts. And I had a couple of guys that kept up with what they were doing, but I didn't get into it.

HT: Did you have a Status of Material and Equipment on Order Report set up by chart of accounts sequence?

90. FM: DuPont had that.

HT: And you kept a cost-and-comparison-to-estimate report?

91. FM: Yes, DuPont did. And I had access to all their reports. I didn't duplicate their staff.

HT: Other than what I just mentioned, did you require reports of DuPont?

92. FM: No, I'd just review their own reports to themselves.

HT: You didn't tie them up making a lot of paperwork to you.

93. FM: That's right. Not to my field office. They had to send all the standard invoices and cost reports, etc. to the Federal Finance Office of course. One thing I did – right from the start – was to have two men from the GAO in Washington, D.C. sent out to Hanford permanently to review all of DuPont's costs.

HT: Now, within your staff, did you have regularly scheduled progress review meetings?

94. FM: Oh no. I didn't have any. I didn't have any standard meetings; I'd call a meeting when something was necessary.

HT: Did you meet with DuPont on a regularly scheduled basis?

95. FM: No, but I talked to DuPont any time I wanted to.

HT: Now, how much time did your staff deal with city, county, and State governments, and with the press?

96. FM: With the press, none. Now on agencies, I established a working relationship with the Governor and several of the wartime agencies – Federal usually – and I got well acquainted with the Congressmen and Senators, and I didn't have any problems with them. The Governor was wonderful – he assigned one of his key people to take care of any problems we had, and I could call him. He didn't know what the project was for.

HT: How about the city of Pasco?

97. FM: The city of Pasco wasn't involved with us, except for our people building lousy camps along the river.

HT: How about Franklin and Benton counties?

98. FM: Benton County. We worked out legal things with them. And when they turned up, we'd find some way to take care of it. It was never a problem. And the Congressman was Hal Holmes – a big help to me.

HT: Switching gears, did DuPont design to a building code?

99. FM: I don't think that anything they did was covered by a building code except some of the camp.

HT: Did you have to deal with the county and the state on building codes? Did their inspectors come in?

100. FM: They didn't come in. But we had a working agreement with them and we followed their rules.

You know, we had a remarkable amount of cooperation, and I think it was the DuPont people who were reasonable, and they didn't try to pull any wise stuff on them, and I had good relationships with them.

HT: Usually that comes about because people like the DuPont people are smart.

101. FM: They were smart. I think we had the best people DuPont could dig up for us.

HT: Did you get many changes of direction from Wilmington?

102. FM: No, their engineering department kept well ahead of the job. I don't know of any changes of that kind.

HT: How about Washington, D.C? They left you pretty much alone, didn't they?

103. FM: Yes. You mean the Army? They left me alone entirely.

HT: Did the Chicago Met Lab make a lot of changes?

104. FM: Not very much. They approved all of the stuff that related to the process. And DuPont wouldn't even build until they had approval.

HT: Did Chicago have representatives at Hanford?

105. FM: No. Chicago would check the design. The Met Lab would come out whenever we needed them, of course.

HT: Were your decisions individual or group decisions?

106. FM: We didn't have any particular system. If I needed a meeting, we had one but most of the time –

Well, for instance, in the middle of '44 Gil Church said to me, "I think we've got to build another bunch of barracks, so we can get some more people. " So I thought about that for a while and I said, "Gil, if you do that you're going to have to hire a lot of people to house the extra people you need, and I think we'd just better extend the working hours to solve our schedule problems, rather than try to get more now." and we discussed it for a while and he said, "Well, I guess you're probably right", so we went ahead and did that. So we settled it in a few minutes.

HT: I recall in McCullough's recent biography of Truman, where his committee member went out to Hanford to try to find out what you were doing. And you told the guard at the gate, "He's not to come in here."

107. FM: The story I got, Truman did not send him there. Truman had an appeal from the Secretary of War to keep his gang away from us. And Truman agreed. And the man who came out here just did it on his own – he had no standing. I wouldn't let him in. He had a motel down in Walla Walla, and I had a security guard go down there and check it out when he tried to get in. So I told him he'd better go back and report to Truman because Truman had agreed not to come here. The guy argued a little, but I told him that he was wasting his time – he wasn't going to get in. And I told him that he'd get a phone call at five AM when it's eight AM in Washington. So I told him he'd better pack up and get ready to go, and I guess that's what happened.

HT: You got there in the middle of March of 1943, and you started camp construction almost at the same time. Who designed your camp?

108. FM: I think DuPont did that, but they had a subcontractor for the camp water system.

HT: Had DuPont gotten a running start on the camp design before you got there?

109. FM: No.

HT: Well, they may have used standard buildings from their other jobs. What kind of water treatment did you have for your camp?

110. FM: There was a little power house that we acquired downstream from where Priest Rapids is now – not too far from where B Reactor is now. That had diverted water from the river into a canal that followed the river and fed all the orchards and stuff clear down a little bit past Hanford. We acquired that plant, and I made a deal for Portland Gas and Electric to manage it, and there we had some conventional water treatment. We had a water facility with an official name, but it was really a swimming pool. And I had an officer on my staff that was pretty sharp at that stuff, so I didn't have to bother with it.

HT: What did you do for sewage treatment?

111. FM: Trickling filters.

HT: When you put up these buildings – the separation buildings and reactors – did you make soil borings first?

112. FM: We made soil borings right after we first picked up the site, and we had the Seattle District Engineer come down and do some drilling, and also do some seismic shots – just enough to confirm the drilled holes. We never did individual drilling for buildings after we had defined the basic nature of the gravels. That valley was gravel to a depth of 50 to 2000 feet.

HT: It wasn't a stratified thing where you might find hidden clay beds. Do you think many people guessed what the plant was all about?

113. FM: No, I don't think so. I knew a lot of workmen speculated, but many of them had worked on DuPont war jobs – munitions plants – and they knew that DuPont had developed a high explosive that was secret for a while, and they just figured DuPont had found a better one.

You know, one time a guy came in to see me – a laborer – and he said he had a radioactive spring somewhere down in the South, and I was shocked. I thought he knew something. But he didn't – he just had a radioactive spring that was supposed to be good for health things – that's what it turned out to be. He just wanted to know if the Government was interested.

HT: Did you have him investigated?

114. FM: Yeah, we did check that one out.

HT: Did security checks for new hires cause you lost time?

115. FM: I don't think so. Most of them were checked out before they came.

HT: You had the top priority. Did the vendors require convincing that you were to get top service because you had priority?

116. FM: No, we had Donald Nelson to do it for us. And we had one of the officers in General Groves office that was dedicated to be sure that we didn't lack something because of priorities.

HT: Did you have expediting problems that were worse than in peace time?

117. FM: I think that we didn't wait for problems to develop. For important items we would put an officer or other expediter in the factory, and he'd get right after them to make sure they took care of our needs. We licked the problem before it started.

HT: Did you have a problem with unskilled labor?

118. FM: I think they were pretty well skilled, if they said they were. I never saw any unskilled labor.

Once we had a bunch of laborers come into my office with a complaint that their lunches weren't very good. And I said, "Alright, you pack 'em yourselves. We'll put out the makings and you do it."

Then they asked about some other crazy things, and finally they said, "We know for a fact that you are serving us a lot of buzzard instead of turkey". And we did use a lot of turkey, because we tried not to use beef, because we had to get points. And I said, "Aw jeez, you guys found us out. We had so many buzzards around to watch you guys and see if you moved or not, that we had to get rid of 'em". And they all thought that was a good joke and it broke up the meeting.

HT: So, you never had a program of training workers - they came in trained.

119. FM: We didn't train them.

HT: This thing about untrained workers I got from one of the books. I wanted to ask if the author got that from you. You remember that story about qualifying a carpenter if his father saw a hammer once – you never said any of that.

120. FM: No, I never heard of it. DuPont did the recruiting, and Gil Church and I made a trip to all of the War Manpower Commission offices and got them oriented that we were important.

HT: A lot of the people that you got were already working in nice places. Did you have to offer them a premium to get them to come to Hanford?

121. FM: No, we had union scales of wages, and that was an average of Seattle and Spokane wages.

HT: Did Dupont's contractors encounter delays in procuring construction equipment?

122. FM: We had a problem with the Corps. A pretty good friend of mine was in charge of equipment for construction for the Army, and when I went out to get a hell of a bunch of cars, and trucks, and all kinds of things, I was supposed to go through him to get the Corps' approval, so I got after him, and he said, "Why, you couldn't possibly need the stuff on this list you're sending me!" And I said, "Look, I'll send you a copy of our project." And I took a map of the whole area – the whole 600 square miles and plotted right in the middle of it the biggest Army munitions plant built, and it was just a little, tiny square in the middle of this big sheet. And I said, "Look, Colonel, this is how big our project is compared with anything else you know about, and we have to have this equipment," and he couldn't believe it. But we got it through the Corps, so I didn't have a problem, really.

HT: Now the difference between the 45,000 peak force in 1944 and the 132,000 total employees through the entire job meant a lot of turnover. Did that delay the job?

123. FM: You know, that isn't really that much over a job of almost three years.

HT: So you couldn't give a number for amount of delay because of that?

124. FM: No.

HT: Well, that also speaks to DuPont's planning. They didn't care who they handed their written instructions to.

125. FM: No, as long as he knew his business.

HT: I was used to 10 to 14 days for rail-transport time from back east in peacetime. How did that compare with what you got?

126. FM: We got it pretty fast, because our guys would get right on it as soon as it was loaded, and they'd keep after the railroads until it got there. So I never heard of its being a problem and I think I would have known about it.

HT: Assuming you had the Manhattan Project a decade earlier, and assuming you had secrecy, but no benefit from wartime authority, how long would you say it would have taken you to do the job?

127. FM: Five or six years. But if it had gone through the normal chain of the scientists doing their development, and then doing the design, and then going to contracted construction, I don't think you could get by with telescoping that into the one job as we did. We did all three things together, starting at the same time. And we had one authority. Now that was another big thing, and one of the main things. We didn't have to get four different agencies, each one wanting a share of the glory and the work. We had one. That made the difference. I want to strongly emphasize that. It was one of the things that made it so successful.

If Vannevar Bush hadn't forced them to put all this stuff together under one control, we would never have got it done in time.

HT: I didn't realize that it was Vannevar Bush that talked Groves into that.

128. FM: He didn't talk Groves into it – he talked President Roosevelt into it! Because the scientists kept saying that they could do it all themselves, and do it all in the normal sequence – scientific development, then the designer, then the contractor – using up time. And we had this telescoped – we did them concurrently. And that was a very, very important thing!

HT: The whole Manhattan Project went in parallel. The thing of three different methods of producing uranium at Oak Ridge was because they didn't know for sure which would work best. Everything went in parallel. And Los Alamos was designing the bomb before Hanford or Oak Ridge had any material to send them.

129. FM: That's right. And that was due to Vannevar Bush's influence right at the start.

HT: You know, it's interesting that a scientist would have that idea.

130. FM: Well, you know he got as many awards in engineering as he did in science. And he was an honorary member of the ASCE, and other engineering societies. He was really good and everyone recognized that here was a high-quality guy.

But, getting back to authority, I had more authority in subcontracting than the Chief of Engineers. I didn't have to go to anybody to get approval.

I was the subject of criticism sometimes. General Groves called me one time when we were in the middle of high pressure, and he said, "You know, I'd like you to set aside one day a week to invite all the construction crew – any of them – to come in and talk to you." And I said, "General, that's a pretty hard thing for me to do – to save out a whole day, just waiting for that. They already know they can come in and see me any time they want, and if I have time, I see 'em." And he argued about it. "You know", he said, "This is something that improved the morale at the Panama Canal so they finally got it built."

I objected again for several reasons, and said that there were several reasons I couldn't take a day out, and I said, "General, you're forgetting one thing – this Hanford job is a big project, and we're doing it in three years, not 15 like the Canal!" And that stopped him cold.

You know, I read David McCullough's "Path Between The Seas", and I figured out that the total man-hours used on the Panama Canal were about the same as were used on Hanford.

HT: You did some pretty big jobs in your time: Kittimat, and COBAST in Brazil, and Baltimore Rapid Transit, and so forth.

What was the best job you ever worked on - the one you enjoyed the most?

131. FM: Hanford. The tension and the challenge and the fear that Germany would do it first. The job was exciting.

HT: Well, I've asked all the questions I can think of. Is there anything else you'd like to comment on?

132. FM: I think we've covered it pretty well.

HT: Same here. I think we're finished. Thanks much Frank. You've been very informative – a very interesting interview.

Although that is the end of Frank Matthias' main commentary, he came up with a number of anecdotes and comments much too interesting to omit, but which didn't fit into the main interview.

Here they are:

Airborne Management

HT: I heard you learned to fly at Hanford.

133. FM: Well, with that large area of work, I could get a lot more done by flying out to the areas.

We had previously decided that we had to patrol the site border to keep out intruders, and with those distances it had to be an air patrol. So we bought a plane and hired a pilot. It turned out that this pilot was also a qualified flying instructor, so I got the idea of being my own pilot, and he taught me. I also went over to the Navy base at Pasco and took some lessons there.

HT: How long did it take you to solo?

134. FM: Three hours.

HT: You had a flight strip at the camp, but where did you land at the areas?

135. FM: On the roads. When we moved the Corps office to Richland at the end of the job I took the plane down there, and I flew back and forth to the areas every day from there.

Once Groves sent his Executive Secretary, Jean O'Leary, out to Hanford to check up on living conditions in the women's camp, so I took her up in my plane for a tour of the site. Well, when she got back to Washington and Groves heard about it, he called me and said, "It's damn lucky for you that you didn't crash, because I sure wouldn't have missed you as much as I would have O'Leary!"

HT: Yeah, she was really valuable back there – helps explain how Groves got along with such a small staff.

Nichols wrote in his book that in July of 1944 he flew into Pendleton and was not looking forward to the two-hour drive to Hanford, so was pleasantly surprised to find you waiting for him with your Piper.

136. FM: Yeah, he did seem pretty pleased at that.

The Comptons Visit Hanford

HT: I was very interested in Arthur Compton's book you loaned me, "Atomic Quest". He was a very unusual man.

137. FM: Oh, he was just a great person! On one of his trips to Hanford he brought his older brother, Karl, and his younger brother, Wilson, with him. Karl was president of M.I.T. at that time and was also involved with the Manhattan Project. The reason Wilson came out to Washington just then was to go to Washington State College, where he had just been appointed president. So I spent a day giving them a tour of the whole project. That was an experience – spending a whole day with the Compton brothers! They were a remarkable family.

Hanford's 25th Anniversary And Johnny Wheeler

138. FM: You know, Mary and I were at the 25th anniversary at Hanford in 1968. They had a big affair. Seaborg and General Groves were there, and Johnny Wheeler was there, and some other scientists, and all the old timers – it was great – gee it was wonderful! Anyway John Wheeler and Mary talked for an hour. Mary thought he was just the most wonderful guy in the world, and I do too. He was very young for being as smart as he was – he was less than 30 years old during the project. Later he spent years trying to figure out gravity.

Fixing the Pentagon Punch List

139. FM: On the Pentagon the early rings would come down to the last five % to complete and that five % would take forever, and Groves asked me to see what I could do to speed up the subsequent rings. So I went to the union business agents and got them to go along with forming a special completion team that would all work together to get it done. So we had one or more men from the carpenters, painters, sheet-metal workers, plumbers, etc. And it worked just fine, and Groves liked it that the work got done so much faster.

Before And After The Fish

140. FM: Colonel Matthias reminisced one afternoon about his life on the family farm in Wisconsin. When he was small, he said, his father's attitude towards him, although kindly, was more than a little remote.

Now Matthias always carried a fishing line and hook around with him then, and one day, when he was about seven, he was crossing a meadowland with his father. They came on this tiny rivulet, perhaps fourteen inches across and a couple of inches deep, and little Franklin said, "Father, let's stop so I can try fishing here." His father protested about the waste of time – "You'll never catch anything in that little trickle."

But he dropped his line in anyway and, to the astonishment of his father, pulled out a fish over a foot and a half long. His father, said Matthias, at once acquired an enormous respect for him.

"His attitude changed overnight," he said, "and ever after that we were quite close. And I've always since thought of my life as either before or after the fish."

Appendix H-2

Interview with

WALTER SIMON

Walter Simon was Dupont's Operations Manager of the Hanford Engineer Works during WW II, and is now retired from the DuPont Corporation.

The interview was held on September 23, 1993 at the Manuscript Section of the Hagley Museum and Library in Wilmington, Delaware.

Those who knew him were saddened to hear of his death in September, 1994.

Appendix H-2

TABLE OF CONTENTS

Chapter No.	Topic	WS Comment No.
5.	The Nature of DuPont	1 -3, 18
5.	Wilmington management	
	Engineering	4, 5, 20, 28
	Wilmington shops	17
5.	Hanford management	
	Simon begins work	8, 9, 26, 27
	Slug canning at Hanford	10, 11
	Pile loading	12, 13
	Manpower management	13, 21, 22, 23, 24, 34
	Management working hours	13, 20
	Parallel activities	14
	Quality Assurance	31, 32
	Living and Working conditions	16, 20, 34
	Clerical work	24
	Travel	25, 26
6.	Met Lab	28
8.	Labor conditions	
	Worker skill	7
	Shifts and hours	33
xx	Germany and the U. S.	15, 16
xx	Bomb materials	16

THE WALTER SIMON INTERVIEW

Harry Thayer (henceforth HT): I have a number of specific questions but first I would be interested in your general observations on Dupont's successful prosecution of the Hanford project.

1. Walter Simon (henceforth WS): If Hanford had come out of a blue sky, that would have been difficult, but DuPont had an organization in being. The war effort started in '41 but before that DuPont Engineering had built smokeless powder plants, TNT plants, Remington arms plants. They had an experienced organization but in '43, when the Hanford project came along, it was starting to run out of things to do. The supply lines which were almost empty in '41 were all full in '43.

HT: You had built all the munitions plants that were needed?

2. WS: More than we needed – they weren't using it as fast. So here was an organization in being of some size which had already used its muscles in building a whole series of plants.

HT: Both Matthias and Groves spoke of the fact that DuPont was extremely proficient – that they did things better than any other contractor they knew of; that supports what you just said about your organization.

3. WS: We had a very strong design division, and we had operating and research people who were accustomed to working with that design division.

Now all we did in nuclear was add the theoretical people in Chicago to the end of that stream. Now Ray Genereaux was in design and he had the feeling of how quickly they could make things very definite and put them on paper and get them out to the people who built them.

HT: With his experience, he could quickly develop the process.

4. WS: Right. But there were many like him. And there was a lot of speed in their work because everyone knew each other well enough that it was handled verbally - everything wasn't written down.

HT: That's an interesting point; Matthias said the same thing. Furthermore, they knew each other well enough, and their performance well enough that they didn't have to write confirming memos to cover themselves.

5. WS: Right. In fact, verbally it moved so fast that it worried some people; they'd ask, "How do we know what's going on?"

HT: It was remarked about the Kaiser shipyards that things were going so fast that by the time a memo might have been delivered, the related task should long since have been done.

6. WS: Exactly.

HT: One of my questions concerns the state of training of the workers you recruited. The majority of opinions that I have read was that if a fellow came in with a union card as an electrician, he knew what he was doing. Have you a comment on this?

7. WS: I think that was the case. The fact that you had to send them so far, the recruiter was judged by the quality of the men he sent. If he sent a man a thousand miles and he couldn't do the job, the recruiter's name would be mud.

You know, one of the humorous things I just thought of was – of course we had to have some kind of entertainment in a town like Richland, and one time they lined up a concert pianist. Well anyhow, this time everyone got off the train in Pasco at two AM, and they started to identify themselves and this fellow said he was a concert pianist, and there was a great uproar – "What are we going to do with a concert pianist?"

HT: When did you get to Hanford?

8. WS: I went out there to stay in April of '44. Before that I went back and forth between Hanford, Chicago, Oak Ridge, and Wilmington. I was accumulating an organization for taking this over. This operation kept in touch with Engineering so that they knew exactly what they were getting, and they made many minor decisions, like whether you could reach a valve or a button.

HT: When were you first assigned to this project?

9. WS: In April of '43.

HT: The slug-canning problem is inconsistently described in the literature. Could you tell about how that problem was solved?

10. WS: Well, for many months they tried to work in the Chicago and DuPont laboratories to find a perfect canned slug with no leaks. But they never could make a perfect one. One of our production men, Earl Swenson, said, "We'll have to start using the best process we have now, and make 500 slugs a day and see what the spread is, and take out the best 10 or 20 slugs that are almost what we need and figure why they are better than the others". And gradually they refined the details of what they were doing, until one day, out of the 500 they had a perfect slug. So they kept up on that approach until they increased the daily yield of perfect slugs. We never got to where you could put 500 slugs through this procedure and get all 500 perfect; you got a bigger and bigger yield – 40 to 50%. I don't think that at the end of the war period that we were running much better than 80 or 90%. Each one had to be tested; you couldn't just guarantee the system. Now all through this there were variations of procedures. There wasn't any major time where you took something and then all the slugs were automatically good.

HT: You don't think that one time late at night, as one of the accounts had it, someone had a sudden flash of inspiration and figured out how to do it?

11. WS: No, that's not the way it worked. It was like the story DuPont used to quote in the old days, that the problem was like the Panama Canal and the slides they had there. They got all the experts down there and they put their heads together, and they said that the solution was to dig! And this was a problem that had to be solved that way – you just had to work on it every day until you got a high enough yield. And there was no sudden burst of creativity. It just gradually got better and better. It wasn't subject to midnight inspiration. So they just about barely had enough cans to charge the first pile when it was ready. It was a close call.

HT: The date I have in my chronology is July of 1944.

12. WS: That's right; the pile had to be loaded in August. The first can was inserted by Fermi – there was quite an audience of both engineering and operations people watching him do this and he turned around and gave a very gracious Italian bow.

HT: Were all three reactors operating in January, or did "F" go critical sometime in February, as DeNeal and Jones state?

13. WS: I would guess that the February date was more correct. There was great pressure to get the first reactor going.

There was less pressure and less manpower on the second, and even less on the third one. We had a full organization, both in engineering and operations for the first one. But the fellows who ran the first one dropped out and that left two-thirds of us for the second one. People don't understand why we didn't do some things at Hanford. You were out in the desert, and if you wanted someone qualified to do something, you had to bring him at least 2000 miles. So if they brought their families out, you had to build a house for them; we never had enough houses. So once you had a certain core there, that was it, and as they got involved and operating there were less of them to bring the next unit into operation, because you weren't going to build another half dozen houses. You had to get along with what you had. DuPont insisted on having housing and having men there with their families.

The men who were there worked all the hours they could work. A ten or twelve hour day was a normal day – there wasn't anything else to do, and they wanted to get it done. Most worked a seven day week.

HT: This brings up a point I meant to ask you. You designed and developed and built by parallel activities, and these activities were in unprecedented fields. Were you ever worried – you and the other DuPont managers – that these parallel activities might not be successful, or that they might not all come together on schedule?

14. WS: All the time. It was a constant worry. You had the reactors and the separation plant which were two entirely different types of operation, with different types of people. And you had the final purification plant and, of course the canning operation. All these had separate groups who trained separately. And having all the timing come out and all the people in place was a constant worry.

Now once May 8 came around – there were two magic dates; one was May 8 and the other was June 15. May 8 was the German surrender, and there was a noticeable relaxation after that. It was Germans that we thought were the target – no one ever thought of using the bomb against Japan – it never occurred to anybody. The European scientists expected Hitler to announce that he had this weapon. They could have done it but they were unwilling to commit the resources. The resources in the United States, I was told at one time, were such that the entire Manhattan Project occupied just two weeks of our entire industrial production for just one year. Now we could afford that but Germany couldn't. This was the only place in the world it could have been done.

HT: I'm beginning to find that out.

15. WS: And it was a peculiar combination of intellectuals – the European scientists who could do it were all over here – the ones outside of Germany – from Italy, Norway, England – they all came here. And we had the industrial background that could handle it. In the whole twentieth century there was never another period when you could have marshalled the same degree of effort. All the various talents – the scientific, the engineering, the production, and the manufacturing – everything, to put it together in that short a time. It could only happen once, and it was really inspired by the intense fear that the Germans were capable.

HT: I remember that Conant made a statement in 1942 that there was great doubt that this type of thing could be done, and secondly he said, we know the Germans are trying it and we think they're ahead of us.

16. WS: They all thought that. Particularly the Europeans – Szilard and Wigner and Fermi – they all thought the Germans had the lead and we were playing catchup.

Now the other date was June 15. There was a great deal of the unknown about plutonium. While the scientists were sure a uranium bomb would go off, they were not sure about a plutonium bomb. So we were given the goal of getting enough material for a test bomb and a full-size bomb. And we met those by June 15. That's all they wanted; they were satisfied with one or two.

We had delivered enough for the Trinity test some days before the test, but we delivered enough for the full-size bomb on the fifteenth. Matthias shipped it down as fast as we got it out, a few grams at a time.

After the fifteenth, there was even more of a letdown. We slowed down the operations and started to streamline things and send people back east. After Japan surrendered we slowed down even further; we barely kept everything turning over. It was the wartime pressure that kept everybody together. Everybody wanted to get back to where they came from. Richland was a very sparse village – no bakery or beauty parlor, one drugstore, one food store, one clothing store – people waited in line for everything. There was only one barber shop; they allotted six minutes for a haircut. They didn't want to bring a lot of barbers in – it would just take up houses they needed for production people.

I think, if I remember the figures correctly, that in a normal city at that time there were ten service people for every one involved in industry, but we had just three service people per production worker.

HT: One thing that impressed me in the Engineering and Procurement History was the great part your Wilmington Shops played in Hanford's design and development. That was one of your great resources.

17. WS: They're closing them down this month. They've decided they no longer need a specialty organization like that. There's so much available service of that type, commercially.

We depended on the shops for our commercial plants too. In the chemical business at that time you couldn't just go out and buy a lot of equipment – you had to make it yourself. They were the backbone in the '20s and '30s of our chemical equipment supply at that time.

HT: Did your construction group date back to the '20s as well?

18. WS: Yes. They came out of WW I; they probably had a small group before that.

HT: Another question that has come to mind is, what plutonium-recovery rate did you get in your separation process?

19. WS: I don't recall the number, but we had a pretty high recovery. We all felt satisfied that end worked better than expected.

The wartime period, as compared with peacetime was a very low-output operation – we were so scared of running into something we didn't know how to handle, we proceeded very cautiously.

HT: Everybody has commented that DuPont was extraordinarily thorough and extraordinarily cautious with the Hanford design and construction – that everything was unknown and that therefore everything had to be followed down to the logical end, and everything had to be developed and tested, and my question is, was the DuPont engineering at Hanford more precisely done than the process plants you had designed before Hanford?

20. WS: One of the reasons I was assigned to Hanford was that before the war I had worked on a number of new processes, such as at Belle, West Virginia. There we were using high pressures to increase reaction rates – ammonia, methanol, and a whole list of chemicals that we were doing under 10,000 pounds pressure. Now where we were doing new things like that, we were just as cautious in commercial work in peace time as we were at Hanford. Things we had done repetitively, such as smokeless powder plants, we did without all that precaution. Whenever we went into a new field – one of the early ones was nitric acid – those original plants were built with great care.

Frank Matthias' engineers were stunned by Dupont's cautious and careful approach to Hanford design.

Groves kept asking why it took us so long to do this or to do that – I got the impression he didn't like us. And after the war he came to an ordinance department meeting at Aberdeen and he came up and threw his arm around me and told everyone what a great guy I was.

They were under pressure and we were under pressure and everyone was tired all the time. Nobody got enough sleep. Because of the three-hour time difference, I had to get up before five so I could check on our current status and then update Wilmington as to what we were doing. Then I'd go back to bed for an hour's sleep. Then I'd go to work. And I'd make the last round of the plant at 11 PM. You know, my six minutes in the barber chair, I slept the whole time.

HT: A number of the workers of that time commented on how well organized Hanford was compared to subsequent projects they worked on, and previous ones for that matter. Does that jibe with what you saw up there?

21. WS: The job, on the whole, was well organized. On the other hand, everything wasn't sweetness and light. Everybody was under tension and everyone was trying to get things done and there was no submerging of conflicts. But conflicts were ultimately settled and no one ended up with mortal grievances against anyone else. Gil Church and I are friends to this day. As things go, it was well organized.

The DuPont forces at Hanford got the cream of the crop from the previous two years of the war effort. All the tried and true experienced people were distilled out of the organization to go to Hanford.

HT: Matthias said the same thing time and again in his interviews – that one of the big reasons for Hanford's success was the tremendous people that DuPont put on the job.

22. WS: For instance, in my organization I had seven men who had been plant managers prior to Hanford, and they headed up different phases like the reactors and the separation plants, and they had all that managerial experience that they applied to these much more subordinate tasks. Hanford was very heavily staffed with competent people.

HT: MacCready, a physical chemist and a nine-year veteran with DuPont was quoted in Sanger as saying, "I had been involved with – startup at about 15 plants, and none of those plants, which were relatively small and simple by comparison, started up as easily and in as trouble-free a fashion as these out here did. I think that was because – they were impressed with the significance of the fact they were going a long ways here on very little information. So everything was most carefully considered as we went along and most carefully checked. – As a consequence there was essentially nothing that had not been accounted for". He was talking about the separation plant there. Would that be roughly what you would say too?

23. WS: Yes. Again, because of the competence and experience of the people who were on it. I know of no other DuPont operation in Dupont's history that was as heavily manned as that one, or about which the board back in Wilmington was more fearful that something would happen that would absolutely destroy the company. Walter Carpenter said time and time again, "You've got the whole company in your hands."

When I was in my thirties those things didn't bother me; now I'd be scared stiff. I can't believe what I went through out there one day at a time.

HT: I'd like to pick up a couple of detail points that I skipped because I didn't know how much time we would have.

I remember reading, especially with the documents I pulled out yesterday, that the clerical work at that time was done with the old-style hand, accounting-card methods, and so forth. Did you use IBM card sorters at either Hanford or Wilmington?

24. WS: They might have used them at Wilmington; I don't remember any out there. Now what we did do – this is so common now that it wouldn't be anything extraordinary – we encouraged all the wives to work.

One of the scientists' wives had never worked before. We worked on her to be the first wife to go into the office - she went into purchasing. She'd never done anything like that in her life. She developed into a first-rate purchasing person. That encouraged all the other women to come in – they were all college-trained women – housewives – but when they went to work on something like that they did an astounding job.

That's the way we did the bulk of our clerical work. We had a large amount of man-and woman-power to take care of these massive routine chores.

But I don't remember anything but hand-cranked calculators.

HT: Another point I'd like to discuss is air travel. A transcontinental flight then took 20 hours, even if everything went right, and often it didn't. Matthias told me that, with two groundings, it took 48 hours from Washington, D.C. to Seattle.

25. WS: He always tried to fly; I never flew. I always took the train, because once I got on the train, they weren't going to put me off. But I could be bumped at any stop on an airplane.

HT: What was it – 4 days to Wilmington?

26. WS: Three days to Chicago and another overnight to Wilmington. You know, in wartime things move fast. I was managing a plant just north of Terre Haute, Indiana – the Wabash River Ordinance Works where we made RDX plastic. I was only 100 miles from Chicago where the DuPont people were starting to try to get the feel of what was going on. These people were engineers and chemists and other scientists and we all knew one another, and they asked me to come up and talk about operating personnel for this plant, but I didn't know what they were making; all they were talking to me about were chemical plants in general.

HT: So you didn't then know about the bomb project. What month was this?

27. WS: Oh, January or February of '43. So gradually I got drawn into this thing. Well, when the concept of this was unfolded, I was absolutely stunned. I came down to Wilmington from Terre Haute to be cleared for this, and they informed me as to what was going on, and I went back to the hotel that night and I couldn't sleep a wink, I was so stimulated. I remember getting up and looking out the window – the streets were deserted, as they were with gas rationing – and wondering, "What is the world coming to?"

All this was at the stage when DuPont was struggling to get a blind man's view of the elephant.

HT: Speaking of Chicago, was that controversy and conflict between DuPont and the Met Lab a pretty real thing, or was it exaggerated?

28. WS: No it wasn't exaggerated; it was a conflict of honest opinion. DuPont, they thought, was making a mountain out of a molehill, and overdoing it.

While we're thinking of it, you asked me over the phone about the extra tubes. Now, that came up in the DuPont Engineering Department first – in design – and then Graves and other people went along with it. But it was Dupont's design philosophy. If you asked them for a one-story office building, they'd always put enough steel in it to put a second floor on it, because they always said, "In a couple of years, you'll ask for a second floor". So that space in the corners of the pile cross section they couldn't see being wasted.

Well, that was the one answer that calmed the Met Lab down; they became quieter.

HT: Did the authors of the Manhattan books interview you?

29. WS: Some did and some didn't. Many of them just called me on the phone; they didn't actually visit me. That's been going on for a number of years.

HT: You know, in all of the published literature nobody has even questioned how it was that DuPont did all this – Dupont's management practices, and I find that very strange. That's what I want to get started – some kind of discussion of that while people like you are still around.

HT: Don Barrie wrote a number of years ago about improvements that could be made in nuclear-power construction. One was to eliminate trivia from QA/QC requirements. Now, QA/QC hadn't been thought of when you did Hanford. What was the reason that a formal quality assurance program wasn't required at Hanford?

30. WS: (A chuckle.) I can't think of the one reason.

HT: Jack Tepe said that it was such a pleasure working at DuPont because you always knew that the fellows around you were doing their job – you didn't have to constantly check on them.

31. WS: That's true – that's true. Now one of my qualifications as a plant manager in the commercial work in the '30s and in the first two years of the war effort I got acquainted with a great many scientific people in DuPont. Somehow or other everyone acquired a valuation of what he could do under a certain circumstance.

So I could assemble an organization of people who largely, in my judgment, and really with very little checking by anybody else, could be in charge of the pile, could be in charge of the separation plant, could be in charge of the canning. They, in turn, selected people that they thought were going to do their job competently. We didn't do this by order of seniority - we did it on the basis of our judgment of an individual's competence.

HT: So once you had organized a group, you knew you had a group that was going to do a job, and you weren't going to have to ask them to sign a form saying they did their job.

32. WS: Right. Their health was good, their intelligence was good, their relationships was good, and they did a job.

It always intrigues me in history – why did Grant make a difference in the Civil War: Because of the people he picked. When he got into an organization, he found who the doers were and who the talkers were, and he told the doers to go ahead and do it.

HT: You appear to have had pretty good productivity at Hanford.

33. WS: One reason was our weekend work. We worked enough people overtime on Sunday so that what the rest of them had to do on Monday got off to a good start. They used the day off to catch up on a lot of little things that would be in the way the next week. They were buying time. The Army said, "Don't hesitate to use money to buy time. Every day may be significant".

HT: Are there any reasons we haven't talked about that explain the efficient job you did at Hanford?

34. WS: The wartime atmosphere, and the isolation. You don't often have a job suspended in mid-air like that. You know, two gallons of gasoline – that's all we got a week. Nobody could run to Seattle for the weekend. Plus the general, underlying wish of everyone there, "Let's get this thing done and get back home".

An element of inefficiency that we had however, we took enough people out there to run three reactors, and they were all there when the first one ran. And two-thirds of them went to the next one. And the final third went to the last reactor with the experience of two startups behind them.

Getting the first reactor going was the most important thing so we overloaded it to be sure we didn't depend on any one person. We were three-deep on that first reactor.

We also had excess people for every one of these, even when we were running all three. But as soon as we passed June 15 we started to thin down the excess to the right amount for each reactor, and sent them back to commercial jobs.

We also had excess people for the separation plants.

Another inefficiency was the very large demineralization plant that we built in case it was needed. It turned out that it wasn't. That was an expensive piece of equipment.

Now you think in terms of today, we had salary and wage controls, and the highest paid man in operations got only $1000 per month for a six-day week. No one went out there to make more money. They went because we convinced them their talents were needed and because they felt they could make a contribution to the war effort. Almost everyone had relatives in the service, and every week someone had someone in their family or that they knew that was killed, so everyone felt very keenly the necessity of what they were doing.

One interesting thing was the Japanese balloons overhead that kept dropping bombs on us. There were 8000 of them that came over; I've seen as many as 40 in one day, and one of them severed the power line from the Priest Rapids sub-station, and that shut us down. When I told Groves about it he said, "Now I suppose you'll want the Purple Heart."

HT: Well Walter, I think we're down to the end, but just one more. You did a lot of good things for DuPont subsequent to 1945, including Manager of the Film Department, and more. What do you think of as the best project you ever worked on?

35 WS: Nothing ever came up to Hanford. After that, everything else was an anti-climax.

HT: Well, many thanks for your time Walter. This has been very interesting and informative.

Appendix H-3

Interview with

RAYMOND P. GENEREAUX

Ray Genereaux was DuPont's Assistant Design Project Manager for separation-plant engineering in the Explosives – TNX Group of the Design Division. He was formerly head of the Chemical Engineering Branch of DuPont's Pure Science Research Group, and the only five-time contributor to the Chemical Engineers' Handbook.

The original interview was conducted by telephone on Sept. 28, 1993. Mr. Genereaux supplied additional information in a number of phone calls, and by letter. He also reviewed the entire manuscript and made a number of helpful comments on it, all of which have been incorporated into the report.

Appendix H-3

TABLE OF CONTENTS

Chapter Outline No.	Topic	RG Comment No.
5	Engineering	
	General remarks	1, 20, 21,
	Communications	20
	Design methods	7-10, 34
	Drawings and checking	13 -16
	Unique design features	20
	Security	29 - 32
	Equipment and vessel tolerances	26
	Pre-project assistance to Met Lab	33
7.2.4	Wilmington Shops	18, 22, 24
7.3	Hanford Management	
	Critical path	2 -6
	Crane operator training	25
	Startup problems	27
—	Miscellaneous conditions	17, 19

THE RAY GENEREAUX INTERVIEW

Harry Thayer (Henceforth HT): Before we get into details, I wonder if you have any general observations of why DuPont was so successful at Hanford.

1. Ray Genereaux (Henceforth RG): DuPont succeeded at Hanford because of teamwork and experience. I also wish to emphasize that DuPont's *normal* engineering and construction practice stressed quality and safety, and one of the main reasons they were so successful in implementing these characteristics was that everyone in design and construction knew each other very well.

And, in the nature of general remarks, let me say that there is no appreciation today of the very great differences between then and now in technologies of all sorts in design and construction

HT: I completely agree, and the differences are far more than marginal. For that reason, I have drafted a section of my report entitled "Technologies of The Forties".

In one technology however, DuPont, Frank Matthias tells me, was a generation *ahead* of the rest of us, and that was in the use of CPM, and that you used it at Hanford.

2. RG: Well, you've come to the right place. One of my men developed it several years prior to Hanford. DuPont first used it on single projects – then applied it to combinations of projects.

HT: Just so we know we're talking the same thing, are we talking nodes, activity lines with durations, longest combination of activities being the critical path, shorter sequences having float, and so forth?

3. RG: That's right, and DuPont gave it to the Navy and they developed PERT from it, which they claimed as their own invention.

HT: After Matthias told me of CPM at Hanford, I talked to Boyd Paulson; he was quite interested that you had originated it that early. The first published reference he knew of was an article in the '50s by a DuPont engineer.

Who managed the CPM for Hanford?

4. RG: The Hanford CPM was organized by the Construction Division. The Design Division went along in parallel with Construction's CPM.

We first took a broad cut at it – then expanded areas on the critical path in greater detail as circumstances required. The broad cut was called the "Master" critical path schedule.

HT: How frequently did you update it?

5. RG: It was continuous.

HT: What was your name for it at the time?

6. RG: We actually called it "Critical Path Scheduling".

HT: Design of those unprecedented systems must have been interesting.

7. RG: It was very exciting. It was going so fast that we talked it out over the drafting boards. We wrote very few memos.

We designed the cell systems to hundredths of an inch, overall. In the field they had a mockup cell in which they put together the complete piping suite to check the fit in the cell. It was then disassembled, transferred to the actual cell, and re-erected. During this fabrication they rectified any discrepancies on the drawings. We had design people in the field to oversee corrections like this. We wanted to keep the systems the way we designed them.

At the start of separation design no one knew what separation process was to be used, so we started off on the canyon cells by designing a "kitchen" in which they could cook anything when they finally decided what the process was. We had standard, remotely installable connections for electricity, gases, water, fluids, etc. so they could later design in anything – mixers, vessels, centrifuges, etc. – anything at all.

I told my engineers never to work with more than three significant figures; two were usually sufficient. With that criterion, we designed the separation plant with six-inch slide rules.

I also told my engineers to design with extreme economy because after Hanford was done we would return to competitive commercial jobs, and I didn't want them to get spoiled for commercial work.

HT: Simon remarked that the painstaking approach you used at Hanford was not unusual. In unusually demanding commercial work, such as the Belle, West Virginia plant which you designed for 1000 atmospheres, you used every bit as much care as you did at Hanford.

8. RG: He's right.

HT: Was it DuPont's practice to field-route piping smaller than two and one half inches, or did you dimension it on the drawings?

9. RG: We put it on the drawings.

HT: Did you have IBM card-sorter machines at Wilmington?

10. RG: I never saw any. All we had besides slide rules were hand-cranked calculators, four-function, that is.

(HT Note: I think it obvious that these calculators would have enabled them to design to 0.01 inch for the cell piping, requiring at least five significant figures.)

HT: How about drawing production?

11. RG: We started off with a drawing list – then added drawings as they became necessary. Then we finished with a "negative drawing list" which was a list of all the items yet to be completed – similar to a construction punch list.

Wilmington's design-and-drawing schedule led the field requirements, and not vice versa. We knew the proper construction sequence – then cranked out the drawings as fast as possible.

HT: Did you issue many preliminary drawings?

12. RG: As few as we had to. When the field asked for them we issued them, of course, except for the separation cells – we never gave out preliminary drawings for them – or very few of them.

HT: I heard that you checked 100% of the work on the drawings.

13. RG: Drawing checking was by a separate checking group.

There was no such thing as spot checking – the checkers ran through all the design steps that the original designers did, so that 100% of the design effort and of the drawings were checked.

Dupont's opinion was that many other engineering firms didn't understand checking.

HT: What were your working hours?

14. RG: The work week was 44 hours. I worked so many hours that I dreamed about it.

HT: Did you have drafting machines?

15. RG: Yes.

HT: How large was your design department?

16. RG: 300 to 500 people.

HT: How did you place long-distance calls?

17. RG: We phoned the operator and gave her the number. We hung up while the operator found a routing for the call. After a while the operator would phone back with the party on the line. Sometimes it took an hour for her to phone back.

HT: I was impressed by the capabilities of Dupont's Wilmington Shops.

18. RG: We used them for the development of cell piping connectors. We also rebuilt Kolmorgen's periscope there, and other items that the manufacturers couldn't design and build according to specifications. We rebuilt the centrifuges there.

HT: How many shifts did you work on your construction jobs in the '30s?

19. RG: Unless there was some crisis, we worked just days.

HT: Would you comment on some of the unique design features you encountered at Hanford?

20. RG: There were critical needs for quality, operability, and continuous operations to a degree far in excess of what had ever been required in any prior process design. In satisfying these critical needs we encountered the following situations and requirements, most of them unprecedented and all of them requiring a degree of refinement never before faced in process design:

- Radiation protection
- Critical mass of the product
- Remote control of the process from behind many feet of concrete shielding
- Remote replacement of equipment and piping
- Flexibility of design to suit a variety of possible processes
- Severe tolerances
- Equipment dependability
- Standardization
- Special equipment joints suitable for remote connection
- Connectors suitable for remote operation
- Mockups for systems pre-assembly and operator training
- Operation with periscopes
- First industrial operation with television
- Inspection of received equipment by disassembly
- Design to minimize moving parts
- Complete and detailed design with no shortcuts whatever

I should say again that we got this done because of the spirit of teamwork among all of us, and because of the excellent communications within design, and among design, operations, and construction. To keep communications going, I made periodic trips to Hanford – two weeks at Hanford, then three weeks back at Wilmington – in repeating cycles. I kept a steamer trunk at Hanford with my things in it.

An example of communications with Construction was the cell mockup at Hanford; Construction didn't want to waste the time required for building it, and they wanted to build the cell piping directly in the cells to save time. I told them, "Absolutely not". We had to have the pre-assembly to ensure that the pipe and equipment suites were built exactly like the drawings to make remote maintenance as simple as possible. After we talked it over, Construction saw my point.

Another example was the need for absolutely plane floors in the cells. When the concrete superintendent said he couldn't do it, I recommended he make and use a stainless-steel screed, machined to hundredths of an inch. He got even more upset about that, but he did it and we got plane floors.

Now to emphasize yet again, Hanford was successful because we did NOT spare detail development and good design in order to save time.

HT: I know of a short-fuse job in 1940 in which the engineers rushed out "bare-bones" drawings to the field to get them started, and then when the pace slacked off in the engineering office, went back to work on the drawings and made them presentable, as a formal record of the job.

21. RG: We didn't work that way at Hanford, or anywhere else. And one thing that kept the project moving was that we were used to making finished drawings in the order of construction sequence. And we maintained a good pace. My standard was 16 days to do a drawing, for commercial work. We couldn't always do that at Hanford because of the very large amount of development work we had to do at the Wilmington Shops and at vendors

HT: DuPont bought the Betts Machine Company in Wilmington in 1917, according to Chandler and Salsbury. Was that the origin of the Wilmington Shops?

22. RG: DuPont had their shops a good while before that. Because of the competitive nature of the powder business, a good deal of DuPont's powder machinery was proprietary, so in order to keep these machines secret, they made them themselves in their shops. That's where the Wilmington Shops came from.

HT: Did DuPont's competition in the chemical industry also have their own shops?

23. RG: I don't know, but I know I've dealt with other engineering firms that did not have their own shops.

HT: Were Wilmington Shops development costs charged to your design budget?

24. RG: Yes, and that was an additional Design cost to the engineering we did on an item before and after the Shops worked out their answers. And the cell mockup at the Shops – the construction of it – was also charged to design, because we had to prove that our cell design was workable before we issued our drawings. The rebuilding of the periscope and its testing, and the centrifuge rebuilding were charged to Design.

The actual fabrication of the connectors, pipe grabbers, impact wrenches, and pipe extractors however, were charged to the Hanford construction accounts via RPG purchase orders.

HT: How about initial training of the Hanford canyon crane operators at Wilmington; was that charged to Design?

25. RG: They weren't trained at Wilmington. I trained them myself at Hanford with the actual canyon cranes, and charged the Hanford construction accounts for the costs.

HT: If your cell systems were designed and built to 100ths of an inch you must have had your vendors build the cell equipment to the same tolerance, either that or accept their standard manufacture – then scale each one and adapt your piping to suit.

26. RG: They built to *our* tolerances. We showed all the required equipment and vessel tolerances on our drawings and the vendors built from the drawings.

HT: MacCready said in Sanger that Separation-Plant startup was practically trouble free. Was that truly the case, or did you actually have some problems?

27. RG: I sure didn't hear of any problems.

HT: Well, you would have been the first to be told, if there were any.

28. RG: I sure would have.

HT: Did arranging your drawings into secure and non-secure types cause lost time?

29. RG: No.

HT: Did you have secure drafting rooms and secure telephones.

30. RG: The drafting rooms were definitely secure, and you had to have a security pass to get into them. One or two of us always went around at night to see that everything was locked up.

We didn't have secure phones, either to Hanford, or the Corps, or to anywhere else. We used very obscure terminology when we talked on the phones.

HT: Like talking about "product" when you meant plutonium.

31. RG: No, we never even used that term on the phone.

HT: Did you have lists of obscure terminology?

32. RG: Not at all. We carried the terms in our heads, after having first discussed it orally with those who would be using them. We operated on a strict need-to-know principle.

HT: In July of 1942 the Met Lab requested that DuPont furnish one of four DuPont men for technical work at Chicago. Between July 24 and Sept. 23, Cooper, Vaughan, and Peery were apparently working at Chicago. Why did Met Lab come to DuPont for technical manpower, did they request manpower from other firms; and what were the disciplines and capabilities of those three DuPont men?

33. RG: I knew Charley Cooper. He was an engineer and was at Chicago when I went out there in October of 1942. He was a good man and he furnished me information for the design of the Separation process. Francis Vaughan was also an engineer, I think and later he went to TNX. I didn't know Peery. I have no knowledge concerning your other questions. Lom Squires would know, and if you phone him say hello for me.

Burns was my assistant when I went out to Chicago, and he later became my counterpart for reactor design under Daniels.

HT: (Mr. Genereaux and I finished with a general conversation, in which he made the following observations:)

34. RG: Concerning parallel activities, you've no doubt observed that the cell design was complete before any input was available from the Clinton SMX unit.

I knew General Groves quite well during the project, and he was a great help. One time he asked if there was anything he could do to help me, and I said, "Yes, keep everyone off my shoulders". By that I meant to keep all the ribbon clerks in the project from slowing down approval of my drawings. So he spoke to a number of people, and my drawings were expedited through Corps approval in their Wilmington office.

And I think one thing that gave everyone in the whole program very great incentive was the German challenge. We all thought that they were working on their bomb as hard as they could, and were most probably ahead of us. After all they had Otto Hahn and Werner Heisenberg on their side! It wasn't until after they surrendered that we found out that they were nowhere nearly as close to a bomb as we had feared.

HT: So you had no choice but to assume that they were really serious competition.

Well, Ray, these talks we've had have been really informative about what you did then, and they've been a real pleasure.

Appendix H-4

Interview with

JOHN B. TEPE

Jack Tepe was a DuPont research chemist assigned to the Met Lab during the plutonium research and development there. He is now retired from DuPont.

The interview was held on September 24, 1993 at the Manuscript Section of the Hagley Museum and Library.

Appendix H-4

TABLE OF CONTENTS

Chapter Outline No.	Topic	JT Comment No.
5.	DuPont Background	12, 16
5.	Wilmington Management Engineering	9, 10
	Caution with commercial work	11
	Causes for Hanford success	12
	Quality Assurance	14, 15, 16
	Wilmington Shops	18
6.	Metallurgical Laboratory	2, 3, 4, 5, 6, 7
7.	The Costs of The Hanford Works	8

THE JOHN B. TEPE INTERVIEW

Harry Thayer (henceforth HT): When did you start work at the Met Lab?

1. Jack Tepe (henceforth JT): The fall of '43. I might add that I worked on atomic research before I went to Chicago in the laboratories here, but not on plutonium. This was service work for some agency that I don't remember. The work involved the separation of isotopes, and on the need-to-know principle I didn't know what the objectives of the project were. It was part of Urey's work at Columbia University. My work on it was in early '43. If you wanted to look for the reports, you would look under the name of Harrison Carlson. We also did work for the Naval Research Laboratory and the Chemical Warfare Service.

HT: Did the Met Lab design the cold end of the separation plant for lanthanum fluoride?

2. JT: We didn't do any design work – just research work. We operated a separation semi-works at the West Stands along with the pile. We used very tall towers in the stair wells for that. I was in research, so I don't know about the separation plant design.

Our main activity at the Met Lab was related to the gaps in the information required by design. Filling these gaps involved the scientists, with engineering assistance.

HT: In the Met Lab's review of drawings, did the scientists attempt detailed checking of the Wilmington drawings?

3. JT: I have no knowledge of those drawings being checked at Chicago in the drafting-room sense. The scientists there checked the drawings for concept. There was nothing like a detailed check. I do know that there was a constant flow of drawings going through Chicago.

HT: I understand that DuPont would not buy or build anything until the Met Lab had signed off on the drawings.

4. JT: That is correct.

HT: I read that the Met Lab worked a seven-day week.

5. JT: I was paid for a six-day week but I worked seven days, and we worked all three shifts.

HT: So, it is likely that most of the scientists worked seven days.

6. JT: Absolutely. There wasn't much correlation between the clock and what was worked. They worked all kinds of crazy hours. Some started at four or five o'clock in the afternoon and worked until four or five o'clock the next morning. They were very individualistic in their work habits. I had operations running at all times, so I would sometimes visit these operations at night.

HT: I'm reminded of the story about Szilard leaving the Met Lab in a huff about something, for a few days, but he wasn't idle; he was hard at work at home.

7. JT: Did you read the recent book about Szilard? It's enlightening. He was in a kind of unusual position. I don't think that he had any line responsibilities that I know of, but he was apparently a companion from the early days, of some of the European scientists, and he was ever present, but it was never apparent to me what his responsibilities were. I was in lots of meetings with these people and I always knew what each guy's responsibilities were. I think you might say he was a consultant, a critic, and a confidant, and probably very helpful in that capacity.

HT: I am in the process of escalating Hanford's 1945 costs to 1993 and I would like your opinion on a question of indices. The equipment indices – Nelson-Farrar and the Chemical Engineering – escalate at a much slower rate than the ENR. One of KE's mining engineers, Jim Thompson, is of the opinion, reasonable I might add, that any equipment manufacturer still in business has had to pay the strictest possible attention to cost cutting in his shop, and in this restricted and controlled operation he has found it possible to do so, a control that the construction industry has found much more difficult. Therefore the ENR has risen more rapidly than the equipment indices.

Have you a comment on this concept?

8. JT: Probably a good answer. The opportunity for controlling labor-intensive activities would permit a slower increase.

HT: I've wondered how Dupont's managers felt about the program of parallel and unprecedented activities. Walter Simon and his associates, he said, were always apprehensive that some of these activities might not succeed, or if they succeeded, they might not be in time to support the rest of the project. Were you worried about that?

9. JT: I think the answer to that question depends on who you talk to. Now Walter Simon was the plant manager. Plant managers are responsible for operating the plant, making quality products at a competitive price, and selling them. I think the scientists fully recognized the risk, but the attitude which I always observed was optimism and confidence and faith that they could solve the problems and it would work. And that applied not just to the nuclear project, but to the commercial work later on. Commercial work involving new processes and new products took similar risks. Management was very nervous because they had the final responsibility, but the technical people, with even greater awareness of the problems and uncertainties, were perhaps more confident because they understood.

Let me say this, that the man who later became my father-in-law – A.W. Skerry – was high in TNX and was a very optimistic person and he was driven on this project by patriotism, and I think his attitude was, "We're going to *make* it work."

HT: What was his function in TNX?

10. JT. TNX was headed by Roger Williams, who was there because he was one of the strongest technical people in DuPont. He was not an administrator. Mr. Skerry was brought over from production of military explosives to be the administrator of TNX under Roger Williams.

HT: Switching gears a little, I asked Simon yesterday if DuPont took greater pains, by far, with Hanford than they did with their normal process design. He said they did, as compared with their normal process design, but when they had a new problem in commercial design, such as their Belle plant in West Virginia in which they designed a process line operating at 15,000 psi, they took every bit as much care with that as they did with Hanford.

11. JT: I would almost agree with that. I worked on the ammonia and other processes at Belle, which operated at 1000 atmospheres, but they were tested at 65,000 pounds.

We cast the reactors at Bethlehem Steel because they were the only place in the country that could do it, and we used hydraulic tests, and we tested to one % strain.

HT: We're all done, but do you have any general comments about why it was that DuPont was so successful at Hanford?

12. JT: Yes, I think it was because they had experience with very sensitive manufacturing operations, not only high pressures but very corrosive and highly toxic processes. They even had radiation experience in the handling of photographic materials. And I do believe that DuPont had a capability from research through construction that was equal to anybody in the world – I'd say the best in the world.

HT: Groves recognized this and he practically demanded that DuPont take over.

13. JT: Yes, he did.

HT: This brings up an allied question, which I asked Simon yesterday: How were you able to achieve quality in the absence of having QA, which we now know is a tremendous burden in getting nuclear stuff done. But you had quality; how did you get quality without having a formal, documented program?

14. JT: We had QA. We didn't call it QA, but we had it. For example, every material we bought was tested to make sure it met the specification. We used enormous quantities of stainless steel and exotic alloys and they were all tested.

HT: You did the essentials of QA without the burden of all the paperwork.

15. JT: That's right, and construction had always had a major inspection capability. They always had inspectors in all the vendors' shops, and when critical machinery was shipped they had inspectors ride the cabooses of freight shipments to make sure they weren't bumped, or lost, so we had QA but we didn't call it that. We had administrative paperwork for all this but it was a minimum, and we didn't talk about it.

There are a lot of things that we're putting labels on that are not new.

HT: Simon had another answer, which was another aspect of it, and he said, "The way I think we solved that problem was that I knew all the people in DuPont, and when I wanted people to do these different jobs, I would pick very highly qualified guys to do them, and that ensured quality because these guys knew what they were doing.

16. JT: He's absolutely right about that. And when DuPont required capability that did not exist inside or outside DuPont, they would send someone to graduate school at an appropriate place to acquire that capability.

And I will say, after almost 40 years at DuPont, that it was a pleasure working there because these various kinds of capability were available, and when a person was assigned responsibility, you could count on him discharging it.

HT: One thing that impressed me after reading your Design And Procurement History several times was that you had a huge resource not available to other AEs in your Wilmington Shops; that bailed you out of all kinds of things.

17. JT: Yes, it did.

HT: Walter said that the reason that DuPont is closing the shops is that all the capabilities of the shop are now available commercially. It doesn't pay to keep them open.

18. JT: That's true, and another reason is that the capabilities are no longer needed. With computer control of manufacturing and advanced technologies we no longer need the shops.

HT: Well, I think we're done, Jack, and I thank you a lot.

Appendix H-5

Interview with

RUSSELL C. STANTON

Russell Stanton was DuPont's Division Engineer in charge of construction of the three 105 Reactors at the Hanford Engineer Works during WW II and was subsequently Assistant Field Project Manager at the AEC's Savannah River Plant. He is now retired from the DuPont Corporation.

The interview was conducted in two telephone calls on October 8 and October 11, 1993, and was supplemented by subsequent phone calls and letters. Mr. Stanton was kind enough to review the Hanford Management section of Chapter 5 and to make numerous and helpful comments on it, all of which have been incorporated into the report.

Appendix H-5

TABLE OF CONTENTS

Chapter No.	Topic	RS Comment No.
5.	The Nature of DuPont	
	Engineer management of crafts	13, 14
	Construction management manual	56
	Construction management training	57, 58, 59
	Engineering	60
5.	Hanford Management	
	Critical path	14, 16, 17
	Engineer management of crafts	1, 2, 3, 4, 9, 12, 13, 42, 43
	Materials expediting	8, 12
	Quality assurance	34 -38
	Inspection	51
	Design changes	55
	Construction – the pile	18 -28, 39, 41, 50
	Construction cost	44, 45
	Hanford chronology	6, 7
8.	Labor Conditions	
	Productivity	29, 30
	Disputes and labor relations	4
	Labor skills	15, 40, 52
	Shifts and hours	48
	Manpower short age	53
10.	Intangibles	14, 29, 30, 31, 60

RUSSELL C. STANTON INTERVIEW

Harry Thayer (Henceforth HT): Did your engineers issue daily written instructions to the craft foremen?

1. Russell Stanton (Henceforth RS): Not written; there wasn't time – everything was moving too fast. The engineers would hold gang-box meetings whenever necessary – sometimes as often as daily – with the craft foremen.

At the start of each new construction phase the engineers would hold a Job-Instruction meeting with the foremen to describe the objectives of that phase.

For occasional, important changes the engineers and secretaries would work at night to issue written material to foremen.

HT: Did the engineers then stay on the job to supervise the crafts?

2. RS: Yes. They were constantly on the job to see that things were going right, and would instruct the craft foremen when necessary.

The engineers also managed the changes that came up. At one time the structural steel for the 105 building was late so, to permit work to continue, we devised a temporary enclosure.

HT: For your engineers to direct the craft foremen, they must have been quite experienced – to know what was a day's work, etc. How old were they, on the average, and had they been with DuPont previously?

3. RS: Well, I was 28, and my engineers were from their mid-30s on up. They had all been with DuPont for quite a while. We had the pick of the crop. Whenever I would think of a man that would fit in well, I would ask for and get him. We had a very good selection of people. When they finished at Hanford, many of them worked up to plant manager.

HT: Did the craft unions object to the engineers directing the foremen as to the work?

4. RS: No. They asked for directions from the engineers and were appreciative of this help they got from the engineers.

During all the time I was at Hanford, I was not conscious of ANY union problems. If there were any union problems, they didn't reflect down to the 105 Area.

HT: When did you get to Hanford, and how long were you there?

5. RS: I was there for less than one and one-half years. I can't remember exactly when I got there, but October or November of 1943.

We lived in some of the farm houses outside of Hanford – the ones that hadn't yet been torn down. It was out in the middle of the desert, and water supply came from one of the original wells.

HT: Had the 105 building been poured by the time you got there?

6. RS: No. Nothing was started. We hadn't even been told exactly where it would be. (This conforms roughly to the Chronology, in which the B Reactor building started on Oct. 10, 1943. Matthias' diary has preliminary 100 B Area construction having already been started on June 10, 1943, so Russell Stanton arrived somewhere between those dates.)

HT: According to some of the dates in the literature, the B Reactor start was in October, but Col Matthias remembers starting B Area excavation – general site grading, presumably – in April, using drawings from Wilmington on which there were no dimensions because design had not progressed that far.

7. RS: Well, he's right about the "no dimensions". I remember that.

HT: Did the engineers direct the activities of the on-site expediters for materials and equipment from warehouse to work site?

8. RS: It wasn't like that. My whole organization took part in that. The clerks and office engineers would receive the receiving reports, and anything on them that concerned my work, they would notify my engineers and the crafts, and we would go get that stuff right away.

I had a coordinator who was excellent at finagling everything we needed. One time there was a last-minute change in which the Met Lab required a vent valve on every one of the 2004 tubes, and this coordinator found them in the 200 Area supplies, and he proceeded to appropriate them for 105. 200 didn't even know they had them so they didn't raise a fuss.

It took a half a day to get the valves and two or three days to install them. On an ordinary job it would have taken three weeks.

HT: Did your engineers direct activities away from the immediate job site, such as at the fabricating shops, pre-erection of scaffolding, the operating engineer superintendent for pre-arrangement of equipment, etc.?

9. RS: Yes. We had several 105-Area shops – the 101 building, the tube shop, the Masonite shop, and so forth. There were also the central shops at White Bluffs. My people would expedite things through these shops.

We had a supply of scaffolding at the 105 building and had it erected whenever we needed it. With respect to heavy equipment, we had a meeting at the start of the job and decided on what equipment would be assigned permanently to 105. If we needed more, we got it from the rest of the job.

HT: How many foremen were there at the 105 Area for the engineers to direct?

10. RS: Well, each superintendent had up to 200 men, and the work gangs were from ten to twenty men each.

HT: Did the rest of the HEW project run by the same engineer-managed system?

11. RS: Yes.

HT: Was Hanford Procurement under control of the engineers, or were they both working from the same schedule?

12. RS: The engineers told procurement what dates they needed what, and procurement got it for them.

We had an expediter for outside orders named Leo Franzell, and he was a big help. He had been in merchandising in Chicago and had a very large family there, and they were all very well connected in Chicago, and he could get us anything we wanted. He would ask, "What kind, and how much, and when do you want it?", and he'd get it for us.

One time he asked the camp director if he wanted any candy.

The director didn't know if he was being kidded, or not, but said "Sure". So Leo said, "How much". So the director, jokingly said, "Oh, about a carload." So Leo got him a carload of candy bars.

HT: Did DuPont manage their other projects by engineering supervision of the crafts, and how long had they been doing it?

13. RS: Yes, this system was prevalent throughout DuPont, and they had been doing it as long as I had been there. DuPont had always had a very close relationship between engineering and the crafts.

HT: Okay, I'm done on the topic of engineer supervision. I would next like to ask you about DuPont's use of CPM scheduling.

I understand from Col. Matthias that DuPont used CPM scheduling at Hanford, and that they had been using it for some time on their previous jobs, but they didn't call it CPM.

Did the managing engineers update the CPM drawings, and how frequently?

14. RS: Yes, we did use it, and we had been using it for at least two or three years. My engineers didn't update it daily – things were moving much too fast for us to take care of that kind of paperwork. Here's how it worked:

We started out with the first issue of the schedule, which was based on Wilmington's design schedule. We would then notice problems with Wilmington's schedule and ask if they could do so-and-so. If they could, they'd accommodate us, but sometimes they couldn't because of lack of design information, or something, so we'd propose something else, and we'd work out the final first issue that way.

But our Field Office always had the initiative in making the schedule.

The central Project office maintained the schedule, but we at 105 were moving so fast we didn't have time to notify them in Central of our changes. We'd carry the changes along in our heads until we came to one that would affect the final completion date. *Then* we'd get with Central and work out what we and the Project as a whole would do, and then the CPM schedule would get updated.

I'll say this, the whole job was very, very well planned out.

We didn't call it CPM at Hanford, but it was CPM all the same. I don't remember exactly, but I think we called it, "Job Schedules".

I would also like to say that it was cooperation that got the Hanford job done. We wouldn't have done it otherwise. DuPont in Wilmington and Hanford and all of our other projects knew each other very well and we all cooperated with each other, and DuPont and the Corps cooperated with each other.

HT: I'd like to go back to the crafts for a minute. If a man came in with a union card saying he was a journeyman, could you count on his knowing his stuff?

15. RS: Yes, generally you could say he was experienced. But you'd know in three weeks if he wasn't, and then he'd be passed out. In the early stages all the incoming workmen were put to the work on the camp, so very few trickled through to work on the project, which was OK because design was not far enough along to give us drawings.

Then when the camp was coming to a close we got all those workers, who were mostly carpenters, and that was OK as long as we had a lot of carpentry to do, but pretty soon we had to put the pressure on our recruiters to get us millwrights and machinists.

HT: Barrie and Paulson in their "Professional Construction Management" state that DuPont's Morgan Walker was the originator of CPM. Did you know him?

16. RS: I knew the name, but I never met him.

HT: Matthias said that DuPont's Frank Mackie wrote the first article on CPM that he saw, and that it came out in the early '60s. Have you read it? I can't find it in the Engineering Digest.

17. RS: Yes, I read it about then, but I don't recall which journal it appeared in.

HT: Now, I'd like to ask about construction of the pile. DuPont's Design and Construction Histories describe your methods in great detail, but the descriptions leave me with a lot of questions.

The blocks were machined to a maximum deviation from true length of .006, but the History states that overall pile dimensions were to be held to .005, measured each way from centerline. Those two statements are not compatible.

Did you carefully select blocks so that the pluses and minuses of length tolerance would add up not to exceed the overall criterion of .005?

18. RS: No. In the first place, the overall tolerance we worked to was more like 1/16 to 1/8, not .005. Then, as far as adding up lengths, we relied on the law of averages, and it always worked out. (Well, with a 1/16 overall tolerance, that's .0625, and five four-foot-long blocks, even if every one was the full .006 over or under would be just .03, so that's well within .0625, and there was no problem.)

HT: Next, the History mentioned the use on B Reactor of piano wire and plumb bobs for overall alignment. That method would be real borderline for 1/16 inch. How did that work out?

19. RS: We used piano wire for aligning some of the tubes, but not for aligning the graphite blocks.

We checked the overall dimensions with an Invar tape to which we fastened a micrometer at one end.

HT: After the trial layup in the 101 building, did you then bore the holes in the blocks?

20. RS: No, before. These holes were precisely aligned for parallelism with the centerline of the block. Then when we laid them down in the reactor building, they lined up.

In the 101 building we checked hole alignment by passing tubes through the holes. Finally, we marked each block with metal stamps to identify its location in the pile.

On the overall pile dimension, you have to realize that the tube-length tolerance was a sixteenth, so there was no point in holding the pile to closer than that.

HT: Next, and finally, I'd like to cover the welding problem of the pile enclosure. The welding conference in Wilmington on Feb. 7, 1944 had decided (a) that welds must be in very short increments, and (b) that after the welding of the increment, the welder would peen that section of weld pass.

What was the length of the increment?

21. RS: About 4 to 6 inches.

HT: Did they chip back to bright metal before the next pass?

22. RS: No.

HT: What tool did they use to peen with?

23. RS: A chipping gun with a blunt, cross-hatched end.

HT: What thickness were these plates, and what style joint?

24. RS: These were the side shields, 1-1/2 or 2 inches thick with vee joints.

HT: Had the welding conference participants had experience with peening for prevention of distortion, or was it theoretical?

25. RS: I can only guess, but I think it was experience.

HT: Where were the side-shield plates made?

26. RS: In the White Bluffs shops.

HT: The history refers to plug welds on the pile enclosure. I'm having a hard time, in the absence of drawings, visualizing where and why you needed plug welds.

27. RS: I know we had some but I'll have to check my records to find out where. (After subsequent reading, I think it likely the plug welds were in the tie straps.)

- End of first phone call -

HT: I forgot to mention that last week I was talking on the phone to a member of DOE-RL's current Site Infrastructure Group, and he remarked that whenever he has to go into the now-abandoned 100 Areas up there, how impressed he is with the quality of the workmanship.

28. RS: That's nice to hear.

HT: You were mentioning on Friday that very nearly at completion of 105 B you got a last-minute requirement from the Met Lab to add a vent valve to each of the 2004 tubes, and that you made the change in three days. What enabled you to do it that fast?

29. RS: Well, the pile was ready to go, it was critical, and we had to get it done. We worked day and night. The pipefitters responded very well – they understood the emergency. We had good craft superintendents and there was a lot of camaraderie between them and the engineers. These were superintendents we had known for years – we had brought them from other jobs where we had known them well and worked with them amicably.

HT: And they didn't resent being pushed extra hard?

30. RS: Not at all. They understood it was an emergency and wanted it done as badly as we did.

Furthermore, they were interested in what they were doing. Everything was so different from what anyone had ever done. All the crafts came up with lots of valuable suggestions – things that had not occurred to the engineers.

HT: Didn't the fact that no one at the craft-worker level knew what they were working on have a deadening effect – they could see no point in what they were doing?

31. RS: I think the mystery of it all had a positive effect – it was intriguing – it piqued their interest.

HT: Did you know what it was all about?

32. RS: Not at first. I was first assigned to the project back in Wilmington in August of 1942, and no one told me anything about the nature of the project. Then one day, this other guy on the project and I had to go down to the Wilmington Shops. We took the trolley car, and on the way he didn't tell me what it was, but he let slip some things which I put with some other things I had heard, so I realized right then what it was – there on the trolley car!

Later, when I got to Hanford, they told me formally what it was.

HT: How many with DuPont at Hanford would you say knew the facts?

33. RS: I would say 50, or even less than that.

HT: I'm interested in how you got the reactors built so quickly, considering the obviously high-quality construction you did and the hundreds of thousands of parts you had to assemble.

34. RS: Well, it keyed on my chief inspector, H.S. Cline. He was responsible for the development of all the special gages and other measuring devices that were required for manufacture, checking and assembly. He saw to it that all these operations were done exactly as specified at the 101 building, the White Bluffs shops, and at the reactors themselves. He kept complete and permanent records of all the manufactured parts and their testing, and the conformance to tests, and he kept them by part name and number.

We had a records chief Reed Kenady, who tracked the receipt and handling of graphite blocks and all the other parts. He'd meet the box cars at the warehouse check off all the items as they were unloaded, record receipt of the item, monitor the warehousing, and record the warehouse location for each piece by name and part number.

After the trial layups in the 101 building, he would secure design instructions for graphite block locations in the pile and prepare work orders for their installation. Previously, he had done work orders for processing the groups of various qualities of blocks, and his final work orders for block installation ensured that the blocks of the right configurations and qualities got installed in exactly the right places.

Complete paper records were kept of all of the above, and the Met Lab would come out periodically to review these records to verify that the pile was being assembled correctly according to the scientific requirements.

HT: What you just described Russ, became known 35 years later as QA/QC with the first issue of ANSI/ASME NQA-1 in 1979. It's interesting that you did it when there were no regulatory requirements then.

35. RS: Well, we thought through the whole construction problem of the pile at the start and decided that we had to do all this to ensure it got done right.

Another problem was to keep the graphite from being contaminated. It was absolutely crucial from the nuclear-physics point of view that the graphite be kept absolutely clean. So we had a complete set of procedures written covering all this. For example, we knew we had to keep all our work clothing absolutely clean to avoid contaminating the graphite from the clothes – the gloves, coveralls, shoe covers, etc. so we had a laundry procedure that specified what soaps and detergents could and could not be used.

HT: Did you also have written procedures for the rest of the pile assembly?

36. RS: We sure did. We got out there before there were drawings to work from, so we couldn't actually build anything yet, so we took the opportunity to plan the entire operation.

HT: How many procedures did you write, altogether?

37. RS: Between two and three hundred.

HT: How many sheets of paper was all that?

38. RS: I would say each procedure averaged over ten pages, so I would say we had well over 3000 pages of procedures. We'd work well into the night; it was a form of recreation because there was really nothing at all to do at night in the first months.

HT: I get different accounts of how many graphite grades there really were.

39. RS: I don't remember now, but it doesn't really matter; we took the drawings and specs and built what they showed.

HT: Were your crafts that arrived at the 105 Area competent, or did you have to train them?

40. RS: We had to train all the people who operated the woodworking machinery we had for machining the graphite blocks in the 101 building. That was totally different work from what anyone had done before.

But otherwise, no. All of the standard crafts knew what they were doing when they came to Hanford.

HT: I want to ask a question that you started on Friday (I'm afraid I interrupted.). How about control of vertical dimensions on block assembly above the base layer? I would think that the plus or minus .005 inches in height of the blocks would give you an irregular base on which to start the next layer – the layer with the first set of tube holes.

41. RS: There was nothing you could really do about that. Everyone figured that the blocks were malleable under the increasing weight of the pile. We tested each layer of holes as we went up by inserting an aluminum tube in the holes, and it always seemed to work.

There was a clearance of .005 to .006 inches between hole ID and tube OD, and the tube was rounded on the insertion end, and we didn't have any problems.

As far as what you mentioned earlier about having to clean up the tube exterior to provide insertion clearance, I don't really understand that. We had a thorough cleaning operation at the tube shop. They were degreased – then stored under contamination-free conditions.

HT: What was the total of your 105-building engineering staff, exclusive of the craft superintendents?

42. RS: In addition to the 10 on the organization chart, the sum of White Bluffs and the 101 building was 25 to 30 engineers, plus 15 to 20 non-technical – clerks and stenographic.

At 105 we had an additional 15 to 20 engineers, plus 15 to 20 non-technical.

HT: I would like to be absolutely sure I understand the exact management relationship between your engineers and the crafts. Given a unit of work requiring a day or two of effort by carpenters, operating engineers, and pipefitters, who issued the orders to get them started and to keep them going on the right track until it was finished?

43. RS: Well, the engineer in charge of that area would have to keep his own notes of what was to happen next, which would include sketches and schedules. It was pretty informal – the engineers and the crafts would see each other all day long, so he would hand a copy of his notes, sketches, and schedule to whomever was in charge – the foreman or the superintendent. The usual medium was the engineer's hand-written notes, and they would describe the work in that area for the next one to three days. Occasionally there would be a reason for typed instructions, but mostly it was the engineer's hand-written sketches and notes.

HT: In your 105 work, were you conscious of doing extra work that you would find uneconomic on a commercial job. I am zeroing in on the statements in DuPont's Histories that extra cost was incurred because of the demands of fast completion.

44. RS: Well, the one occasion that I mentioned of having to provide a temporary housing around the reactor erection because drawings and materials were not available to build the permanent 105 structure. That's one example.

HT: Was that type of work common in your area, or unusual?

45. RS: Very unusual.

HT: There are three versions of the solution to the slug-canning problem. Do you know how it was solved?

46. RS: I knew very little about it. That was in Operations under Walter Simon, and he took care of it.

HT: Well, his account was one of the three versions, so if he was running it, I would have to say that his account was correct.

47. RS: I agree.

HT: You worked two nines per day, six days per week. How many hours did you personally work?

48. RS: If I got four hours sleep a night, I was doing well, so that would make it a 20 hour day.

HT: That confirms what Walter Simon said, that they were always tired. Whenever he went in for a haircut he would sleep through the whole thing.

49. RS: That reminds me of how I would get a haircut. The barber shop had fifty barbers lined up in a row. So I would go down with my machinist superintendent and let him go first.

He would sit and watch the barbers and see how they handled their barber tools, and after a while he would say, "Now that one is a real mechanic, and he'd get that one to cut his hair, and wait for him and have him do me next, and I always got a good haircut that way.

HT: Getting into the famous 1500 vs. 2000 tubes question, supposing the Met Lab had prevailed and B had been built with 1500 tubes, how long after it failed would you have required to start all over and build a new one with 2000 tubes – after you received the drawings, that is?

50. RS: About eight to ten months.

HT: Myers had 79 inspectors for the HEW as a whole; did he inspect the 105?

51. RS: No, we did all our own inspection under Cline. We had very specialized work.

HT: The Construction History has a detailed account of the painstaking search for welders for the side shields, the tests given these highly qualified men, and the fact that less than 20% of them could pass the special Hanford, 105 welding tests. What do you know about that?

52. RS: I don't think that happened. I never had any shortage of qualified welders. We'd get 'em in and give them two or three days to show them what we wanted and they did fine.

HT: There are many accounts in the Construction History about delays because of lack of manpower. Did that occur in the 105 Buildings?

53. RS: I can not think of a single case of schedule slippage due to lack of manpower at the 105 building.

We did have to make sure that we planned around the available drawings, materials, and manpower however. In other words, we made sure by looking ahead and planning for it, that we never got caught short. [1]

HT: Did the coal-fired plants run continuously on spinning reserve for cut-in on Bonneville failure, or were they part of permanent pumping by being hooked up to some of the pumps?

54. RS: The former.

HT: Did you get a significant number of design changes from Wilmington?

55. RS: We did get a number of design changes, but they were for detail items, and not for major rework. They were annoying but not critical. It was not a major problem.

HT: Did Dupont's Construction Department in 1942 have its own manual of construction management?

56. RS: Well, not in those words, no. There were quite a number of rules and regulations – sheets and sets of paper, but an instructional manual, no. That didn't start until after the war.

HT: Did the DuPont Construction Department give a construction management training course to its promising young engineers?

57. RS: I suppose so, but it was so informal, I never knew I was in it. After several years I happened to see a brochure that DuPont sent out to colleges to recruit people, and my picture was on the cover, so I figured I must have been fairly well regarded.

HT: Are you saying that they put you to work and all the while the old hands would say, "Do this", and "Don't do that again", and so forth – on-the-job training?

58. RS: Exactly.

HT: Prior to 1943 did you ever see a book on construction management?

59. RS: No.

HT: Beyond cooperation within DuPont, and with the Corps, and having the pick of DuPont people, are there any other reasons for the Hanford success – questions I should have asked but didn't?

60. RS: Well, those reasons are good, but beyond that – it's hard to say. What we did was just a matter of course then.

Much later, when I was Assistant Field Project Manager at our Savannah River project, which had three reactors, it went pretty well, but still it took almost three years, and we'd sit around asking, "Why so long?" "It wasn't this bad at Hanford". In fact, I guess everyone else down there got pretty tired of hearing about Hanford.

It was a different time.

As to something more specific, there was a definite incentive in the national emergency – that made a difference. Also we had total support from the Corps in Washington, D.C.

Another thing was that DuPont had a tremendous engineering organization.

We had internationally renowned experts in every phase of engineering. We would have some question in the field as to what kind of material to use, and we'd

---

1 Mr. Stanton is essentially correct. The construction progress charts of DuPont B show that overrun of the three 105 building schedules are relatively minor; 105-B – 2 weeks (4.7% of 43 wk. Sched.); 105-D – 5 1/2 wks. (13.8% of 40 wks.); 105-F – 6 1/2 wks. (16.3%)

phone Wilmington. "No, don't use that", they'd say, "But I'll have to check". And it wouldn't be more than five minutes and they'd phone back and tell us what we should do.

HT: Well, Russ, that's finally it. I've enjoyed these conversations and thanks very much.

INDEX

This is an index of topics in chapters 1 through 10. Indexing of chapters 11 and 12 was not thought necessary because these chapters are a summary and overview of the discussions in the previous chapters.

The appendices are not indexed here, but appendices G and H are provided with their own indices, and tables and figures are listed and located in Appendix A.

ANSI/ASME NQA-1 73
Areas 100B 5, 6, 8; 100D and 100F 5, 6
Areas 200 East, 200 West, and 200 North 6, 11
Attitudes of workers 89, 90, 93
Auditors' reduction of DuPont fee 84
Authority 21, 25, 27; and responsibility 25

Beck, David 89
Bellows, expansion 7
Boredom 99
B Reactor 10
Building Trades Council of Pasco 89
Bush, Vannevar 3, 21

Cans and canning 7, 10, 20, 42
Canyon (221 Building) 11, 13, 14
Carpenter, Walter P. 31, 41, 73, 79, 84
Caution in engineering 45, 46
Cell, process 11, 13, 15, 19, 42, 50–52, 73
Centrifuge 15, 51, 54
Chain reaction 5, 10
Charging machine 7
Checking 47
Chemical Engineers' Handbook 2, 31
Church, Gilbert P. 29, 38, 39, 52, 63
Command channels 25
Clinton Engineer Works 3, 42, 45, 54, 70, 75, 76, 79
CMX 10
Columbia River 5, 6, 10
Competence of crafts 89, 90

Competing processes, construction of 18
Compton, Arthur H. 53, 77, 78, 98, 108
Cooling water 5, 7
Conant, James B. 18, 21
Concentration building 11, 14
Congressional oversight 26
Construction subsidiary to Engineering 35
Construction (1940s) 95
Control rods 7
Cooper, Charles M. 30, 80
Cooperation 23, 27, 29
Cooperation by industry 98, 99
Corps of Engineers 21, 26, 27
Cost escalation 81, 83, 84
Cost, management and engineering 82, 86, 87
Cost vs. duration of project 87, 88
Costs, Manhattan, relative to U.S. and German economies 84, 86
Costs, recorded 81
Crane, overhead, canyon 11, 51, 52
Critical Path Method 66–68; DuPont's invention of 66, 67; use at Hanford 67

"Day's Pay" 99, 100
Decontamination 11
Defense acquisition policy 27
Design changes 53, 54
Design-development charges 48, 49
Drawing list and schedule 48
Drawings extremely detailed 47
Dummy slugs 5

DuPont 1, 2, 3, 10, 18, 19, 26, 27, 29, 30–74, 79, 80, 82, 84, 86, 88, 90, 93, 99; Construction Division 35; Design Division 35; Engineering Dept. 35, 37; Explosives TNX 35; general organization at Wilmington 35–37; nature of 31–35; pre-Hanford engineering, construction, and veterans' comments on 31–35; research 31; and turnkey management 38
Duration of projects vs. costs 87, 88

Engineering, DuPont's for Hanford 41–54
Engineering perspective 41
Engineering practices, DuPont's 45–50
Engineering methods (1940s) 96
Engineer supervision of crafts 68, 69
Escalated costs 84
Expediting 27
Explosives TNX 2, 3, 4, 10, 35, 36, 41, 50

Fee, DuPont's for Hanford 84
Fear of Germany 18, 19, 98
Fermi, Enrico 10, 76, 79
Field management, Hanford 63–74
Financial control, Manhattan 25, 27
Fission products 5, 10
Franck, James 76
Fuel tubes 5, 7, 9, 20

Gas rationing 99, 101
Genereaux, Raymond P. 2, 31, 34, 38, 39, 42, 45, 47, 48, 50, 51, 52, 54, 67, 98
Gerber, Michelle 47
Governmental relations 27
Graphite and graphite blocks 5, 7, 10, 19, 20
Graves, George 10, 31, 34, 47, 53
Greenwalt, Crawford 10, 31, 40, 47, 53, 80; Dec. 2, 1942 notes by 33
Gross national product, 1944 86
Groves, Maj. Gen. Leslie 18, 21, 24, 25, 26, 29, 30, 34, 48, 84, 98, 99, 108
Gun barrel tube 7

Hires, total 93
Hours and shifts 89, 93

Immediate start of design and construction 41
Industry cooperation 98, 99

Intangibles, negative 99
Iodine 131 5

Jumpers 15, 51
Jurisdictional boundaries 89
Jurisdictional walkout 90

K value 20
Kaiser engineers 2
Kaiser, Henry 45, 108
Kennedy, Joseph 75
"Kitchen," as bounding design for separation cell 50
Kiwanis Club (Pasco) letter 84

Labor conditions 89–94
Legal controls 26
Lewis, W. K. 2, 33
Long distance telephone (1940s) 97
Los Alamos 3, 4, 75

Mackie, Frank 53
Magazine storage 11
Management, DuPont; engineering 41–54; field organization 63–74; general organization at Wilmington 35–37; procurement 55–63; turnkey 38
Management of labor 89–94
Management, Manhattan District 21–26
Management, OIC, Hanford 21, 23, 26–29
Manhattan Engineer District 1, 3, 21, 26, 27, 45
Manhattan cost relative to U.S. and German economies 84, 86
Manhattan Project 1, 2, 81
Masonite 7
Massachusetts Institute of Technology 2, 30
Mathias, Col. Franklin T. 2, 21, 24, 26, 27, 29, 45, 53, 66, 68, 69, 73, 78, 84, 89, 90, 97, 98, 99
MED activities and attributes 25, 26
MED's Hanford organization 26–29
Meetings 27, 29
Metallurgical Laboratory 2, 3, 10, 18, 30, 41, 50, 51, 53, 68, 70; primary goals 75; principal tasks 75; development schedule 76, 79; DuPont relationship 79, 80
Military Policy Committee 3, 21
Mockups 19, 51, 52
Motivation, management 98

National Defense Research Committee 3, 21

Nature of DuPont 31–35
Neptunium 5
Neutrons 5
Nichols, Col. Kenneth D. 21, 25, 26, 27, 29, 99
No-strike pledge 89
Nozzle 7
NQA-1 73

Office of Scientific Research and Development 3, 21
Officer In Charge 2, 29
On-the-spot decisions 25
Operationally realistic design 47
Organization, simple and direct 25

Parallel and rational schedules 45, 46
Parallel design and construction 41–45, 50, 51
Parker, James 90, 99, 102
Pasco 6
Patriotism, crafts 89, 99
Periscopes 52, 54
Perry, John 31
Phased construction 42
Pile 5, 7, 10
Planning decisions, basic 41
Plutonium 1, 2, 5, 11, 18, 19; pilot production reactor 3
Poison 10
Pool, storage 5, 9, 11
Priority 25
Process cell; see Cell, Process
Process instrumentation (1940s) 96
Procurement 27, 55–63; principal time-saving modifications 55; groups, ad hoc 55; stainless plate warehousing 63, 88
Procurement, separation plant 56
Project cost vs. duration 87, 88
Purification building 11

Quality assurance 69–73; overall 69; reactor QA-preparation 70, 71; reactor QA-construction 72; separation plant QA 73
Quality of Hanford management 73, 74

Reactor building 8
Reactor design 52, 53; principal tasks 53; who advocated 2004 tubes? 53; consequences of a failed 1500-tube reactor 53
Read, Granville M. 39
Recruiting of labor 89, 93
Remote erection 52
Remote maintenance and operation 11, 19; design for 51
Reybold, General 21
Richland 6, 11, 99, 100, 102

S-1 Committee 3
Safety 93
Seaborg, Prof. Glenn T. 2, 18, 31, 53, 73, 74, 75–80, 97, 98
Secretary of War 3
Security design 48
Segre, Emilio 75
Separation cell, remote assembly 52
Separation plant, physical design 50; basic engineering 50; detailed implementation 51; precision of 51
Separation process and plant 2, 5, 11, 12, 19; pilot plant 3
Separation process conceptual design 75–77; incomplete data 50; known data 50
Shield, biological 7, 19
Shield, thermal 7
Shifts and hours 89, 93
Shortage of crafts 89
Significant figures, design 48
Simon, Walter O. 2, 34, 35, 42, 45, 48, 54, 66, 73, 79, 80, 93, 97, 98, 99
Simple designs 47
Single authority 21, 23
Site approval, sole authority 26
Site investigation 26
Size of HEW 16, 17
Sleeve 7
Slugs 5, 7, 10, 20
Small staff 25
SMX plant 3, 10, 70
Speed of design 48
Speer, Albert 86
Squires, Lombard 2, 38, 47, 76, 80, 99
Stainless plate warehousing 63, 88
Stanton, Russell C. 2, 27, 35, 39, 53, 54, 66, 67, 68, 69, 70, 72, 73, 80, 90, 93, 98
Sunday work 93
Surveying (1940s) 95

Teamsters, Western Conference 89
Television 52, 54

Tepe, John B. 2, 35, 42, 54, 69, 73
Time-saving decisions 88
TNX; see Explosives TNX
Transportation (1940s) 97
Trust 29
Turnkey management 38

U 235 3
University of California 2, 75, 76
University of Chicago 3, 75
Uranium 5
U.S. Mail (1940s) 97

Verbal communications 25, 27, 29, 34, 48

Wahl, Arthur 75
Waste storage 5
Welch, William 99, 102
Wheeler, John 10, 53
White Bluffs shops 6, 69
Wigner, Eugene 76, 98
Williams, Roger 34, 35, 38
Wilmington shops 54, 56, 62
Work stoppages 89, 90

Xenon 10